Routledge R

Science and Hu

Originally delivered as a series of lectures for the Halley Stewart Trust in 1926, Lodge's work was collected and first published in 1927. Lodge uses his scientific training to inquire into such general issues as religion, human progress, and societal advances with an aim to better understand the physical order of the universe. This title will be of interest to students of philosophy, particularly those interested in the development of early twentieth century thought.

Science and Human Progress

Sir Oliver Lodge

First published in 1927
by Methuen & Co., Ltd.

This edition first published in 2016 by Routledge
2 Park Square, Milton Park, Abingdon, Oxon, OX14 4RN
and by Routledge
711 Third Avenue, New York, NY 10017

Routledge is an imprint of the Taylor & Francis Group, an informa business

© 1927 Sir Oliver Lodge

Publisher's Note
The publisher has gone to great lengths to ensure the quality of this
reprint but points out that some imperfections in the original copies may
be apparent.

Disclaimer
The publisher has made every effort to trace copyright holders and
welcomes correspondence from those they have been unable to contact.

A Library of Congress record exists under LC control number: 27015718

ISBN 13: 978-1-138-19251-5 (hbk)
ISBN 13: 978-1-315-63989-5 (ebk)
ISBN 13: 978-1-138-19252-2 (pbk)

HALLEY STEWART LECTURES, 1926

✳

SCIENCE
AND HUMAN PROGRESS

BY

SIR OLIVER LODGE

LONDON
GEORGE ALLEN & UNWIN LTD.
MUSEUM STREET

First published . April 1927
Reprinted   . .   June 1927

Printed in Great Britain by
Unwin Brothers, Ltd., Woking

PREFATORY NOTE

THE Trustees cannot allow this first volume issued under the auspices of the Trust to go to press without a grateful acknowledgment to Sir Oliver Lodge of the service he has rendered by the delivery and publication of this course of Lectures on Science and Human Progress—service to the Trustees by his keen appreciation and elucidation of the far-reaching object of the Trust, the enhancement of human well-being and well-doing, and service to the large and entranced audiences that listened to the Lectures as he brought to view the rich contribution that Science can render to the cause of Human Progress. The Trustees look forward with confidence to the enjoyment that will be experienced by the wide circle of readers who will share their pleasure in following Sir Oliver Lodge in the realm of Science in his quest of the main object of the Trust—Research for the Christian Ideal in all Social Life. The Trustees also take pleasure in confirming the accuracy of the view expressed by Sir Oliver Lodge in the opening Lecture that the Founder of the Trust designed that it should be "free and open and unhampered by restriction such as may at any time become out of date—a living Trust, ready to be adapted and utilised in accordance with the necessities of the time."

On behalf of the Trustees,

HALLEY STEWART,

Chairman.

18th January, 1927.

CONTENTS

Science and Human Progress

KNOWLEDGE AND PROGRESS

A GENERAL SURVEY OF MAN'S POSITION AND OF
HIS ADVANCES IN THE NINETEENTH CENTURY.

THIS Inaugural Course of a series of lectures,
initiated and endowed by Mr. Halley Stewart, on
the general theme of Religion, Social Betterment,
and Human Progress, has been entrusted to
me because I regard the Universe from a point
of view attained through a training in scientific
study, and because I have from time to time
addressed audiences on aspects of the universe
that are more or less in harmony with the
intentions of the Founder and the scheme of the
Trust. I understand the Trust to be free and
open and unhampered by restriction such as may
at any time become out of date—a living Trust,
ready to be adapted and utilised in accordance
with the necessities of the time.

In carrying out a portion of the scheme this
year, I agree that I have a message to deliver ;

I have, indeed, already partially delivered it on various occasions ; and I propose now formally to express the substance of what from time to time I have orally said, so as to record it and make it accessible to a much wider circle.   For I am impressed with the majesty and possibilities of the Universe, as contrasted with the comparatively narrow outlook taken by the average of those engaged in the work of the world.   The world has a long, long time before it, and I am convinced that the spirit of man is capable of much higher development in the future than it has yet attained.

There is room for improvement.   The standard of corporate attainment is not high.   Inequalities are rampant.   The struggle for existence is no mere biological phrase ; it is still sometimes a fierce reality.   We are not often aware of it, but sometimes it is brought home even to the thoughtless and the preoccupied.   Comfort protects us from many crude temptations, but the savage snarl might break out in any of us if food and comfort were denied.   Here is an incident which occurred soon after Christmas Day, A.D. 1926. Gruesome it is ; some will think it barbarous ; more, I hope, will think it pitiful.   It happened in Paris : it might have happened in any city.

In the piercing cold of Tuesday night [December 28, 1926], when the canal was packed with ice-floes, two half-starved and shivering members of the city's underworld fought to the death for shelter from the biting wind beneath a strip of tarpaulin. It is used in the daytime by stone-breakers.

The fight was witnessed by some twenty other homeless dregs of society, who come every night shivering in their rags to make their beds on sacks and heaps of rubbish behind casks and other packages which litter the bank of the canal and afford some shelter from the piercing cold.

Nobody thought of interfering until one of the two, known to his comrades in distress merely as " Georges," fell stabbed through the heart. The other then wiped and put away his knife, and crept beneath the shelter of the tarpaulin for which he had fought.

Alas for Georges, and for the winner of the tarpaulin, and for the dismal, reprehensible responsible squalor which made such doings possible ! Dare we claim immunity, if circumstances reduced our condition to that level ? Is any nation immune ? Surely there have been murderous struggles over disputed territory, with no excuse so valid as dire necessity. We cast no stone, but we sorrowfully recognise that while such things can go on, whether on a small or a large scale, our corporate standard is woefully low, we are not truly civilised, man is not Man as yet.

Instinctively we all desire that man should rise in the scale of existence ; for even though our conscious aim may seem to be only improvement in our own individual lot, yet happiness is catching and benefits are shared. Moreover, no individual can flourish happily if the society around him is rotten, so it is only reasonable that each should desire and work for the advancement of mankind as a whole. Towards that great object many of the paths must be indirect, though some are more direct than others. There is a path through

Business and Commerce, when based upon straightforward and honest intercourse between man and man. There is a path through manual Labour, with its opportunities for meditation, and for corporate action with one's fellows. There is a path through Literature and Poetry and Art generally, wherein the human imagination takes its highest flights. There is a path through History, with its study of the past and its application of the lessons to the problems of to-day. There is a path through Politics and Statesmanship, which, freed from party contests, would seem the most direct of all towards the social betterment of man. And there is a path through Religion, which cultivates the highest welfare of the individual, and seeks to concentrate his affection on things above, and not on things of the earth.

Compared with these, the path through Science, with its exploration of the more material aspects of the universe, must seem very indirect. Yet increase of knowledge is essential, and we little know whither that increase of knowledge will lead. Indirect methods are often the most effective in the long run. Mrs. Sidney Webb once told me a saying of her distinguished friend, Herbert Spencer; one of his mottoes was, "Don't hammer on the bulge." By this he meant that if you want to get rid of a bulge or swelling in a sheet of metal, and you hammer on it, you will merely send it elsewhere; but if you hammer everywhere else, and so influence the

whole surroundings, the defect will subside
and disappear of itself. Evils are often best
attacked, and good results best attained, by a
policy of Indirection. Direct prohibition, direct
action of any kind, is a policy of doubtful wisdom.

For myself I do not feel oppressed by any
conflict between Religion and Science, when
both are reasonably understood. Both involve
knowledge of certain aspects of one and the same
Universe ; and controversies arising between
them must be of the nature of misunderstandings
and mistakes, mainly due to our looking at one
aspect from the point of view of the other. Men
of eminence are striving to avoid this mis-
take. The recent book, *Holism and Evolution*, by
General the Rt. Hon. J. C. Smuts, whether we
agree with it or not, is a sign of the times.

Continually I am impressed with the utter
absence of infallibility in human beings, whether
it be about the subject-matter of science or anything else. It has been said that those who make
no mistakes make nothing. However that may
be, it is undoubtedly human to err, even with
the best intentions ; it is easy to grasp at truth
by the wrong end, to emphasise one aspect to
the exclusion of all others, and in general, by
intensity of concentration in one direction, to
narrow down our outlook and close our eyes to
the mystery and magnificence beyond. This
limitation of outlook occurs in men of science
no less than in any other department of human
activity. It has occurred among Theologians too.

They have a difficult though noble task. It is
no light matter, it must always be rather presumptuous, to interpret Divine Realities into human
symbols, and, in Milton's phrase, to seek to
justify the ways of God to man.

Physical science is what I have myself chiefly
studied, but not exclusively ; I have been interested in psychical science too. There is indeed
no sharp line of demarcation between them ;
everything is linked with everything else. We
may, however, choose our own range of study :
we may concentrate on a certain small patch and
penetrate deep, or we can take a broad survey.
By the one method there is a limitation in breadth ;
by the other method we run the risk of being
superficial. To cultivate both breadth and depth
can only be the prerogative of a few. If both
breadth and depth are attended to, by the
same individual, they must be attended to at
different times : it is not possible to work them
concurrently.

I take it that in this course of lectures it is
breadth rather than depth that is most appropriate. There are other occasions when it would
be suitable to seek to penetrate the secrets of the
atom or the ether. Here we can only take a
broad survey ; bearing in mind that every
perceived fact, every reasonable speculation, and
every advance in knowledge, must have a bearing
one way or another on the cultivation of the
human spirit and the progress and destiny of
man.

A Survey.

Let us, then, first take a survey of the Universe from the point of view of modern science, and try to realise what sort of place it is in which our consciousness has wakened and in which we find ourselves ; each trying to do his duty and carry on the work which lies ready to his hand ; hoping, as many must hope, that there is some deep meaning in it all, and that we are in the hands of a Higher Power Who understands the reason of all the manifold activities we see around us, and Who can already perceive—even though it be far ahead—the object that will ultimately be attained by our own individual striving and effort and pain.

We must understand that this Universe is a most comprehensive thing. It includes everything that is ; what we call good and evil, joy and pain, beauty and ugliness, health and disease, every kind of contrast : nothing that exists is outside it. We cannot compare this Universe with any other, or pass any judgment upon it. There is nothing to compare it with : there is only one Universe ; it is unique.

This fact is often a great puzzle to our small intelligences. There are things in existence of which we disapprove. Strange that that should be possible ! Strange that a creature should venture either to approve or disapprove Reality. Yet it seems inevitable that we do ; and the problems so raised are rather inscrutable. Some things excite our admiration, others our disgust. Many have been the attempts of philosophers and theologians to explain these things. For the ultimate solution we shall have to wait.

Meanwhile I expect that any disapproval or dislike that we may feel towards the nature of things, or even of one another, may be based on imperfect knowledge, on some temporary misapprehension. If we could really shatter things to bits and mould them nearer to our heart's desire, what a mess we should make! Our capacity for understanding is very limited; we are puzzled by many mysteries. We are afflicted with the problem of evil : and indeed we feel that against cruelty, oppression, and selfishness the Heavens themselves rebel.

But apart from such things as cruelty and selfishness, I am sure the less we judge the better. The old system of judicial torment and reckless punishment has gone, never to return. There is a sense of greater corporate responsibility. The old rule of eye for eye and tooth for tooth has been superseded by a far higher law. The proverb says that to understand everything is to forgive everything. In the new System, inaugurated 1900 years ago, love and forgiveness are the dominant features. We are encouraged not to judge or to condemn, but to sympathise and understand. Depend upon it that humanity, unless utterly warped by greed and case-hardened by selfishness, will sooner or later respond to this new treatment. The attempt has not yet been widely made : it has been made here and there by individuals, and occasionally by a special Society. But into the work of the world loving-kindness is now beginning to enter : its possibilities are being realised more clearly than ever before : it is beginning, though only beginning, to penetrate every walk of life. There will come a day when human intercourse will be saturated with it, and when the Mind of Christ will be realised and supreme.

To contribute towards that end, to aid in the coming of the Kingdom in however small a way, was the object of the founder of these lectures : and in one way or another it is the object of many statesmen and others who are now in the eye of the world, but who seldom find time to formulate, perhaps not even to themselves, their end and aim. The coming of the Kingdom must be identical with the reign of love. Perfect happiness is incompatible with anything but love and good will. Love manifests itself in many ways, and all its ways are good. It is a thing deep-seated in the Universe, and its manifestations are all around us. It is the absence of love that is deadly ; it is things done without love that are evil. This is what the Master insisted on continually and in many forms. Even sins can be forgiven to those who love much. And again, "Love is the fulfilling of the law"; it is essential to the joy of existence. It is, I believe, an essential ingredient in Deity Itself. The very existence of our powers and faculties shows that it must be so.   The joy of Creation must be of this nature ; and in its doings, in its love-inspired activities, as Browning says, " God renews His ancient rapture."

It is difficult to suppose that much rapture can be felt by any Higher Being about the status and present doings of mankind.   Man is a remarkably imperfect being, and only by taking a very large and far-seeing view can he be regarded as a promising production.   His advent —which in the long course of evolution has been comparatively recent—was probably an exceedingly important event, towards which for millions of years this planet earth had been preparing ; but, as we are still in only the early stages of the process of human development, we are more impressed with the imperfection—with the pain and rebellion and sorrow and strife— than with the joy of a beautifully finished work of art.   The Dean of St. Paul's has recently emphasised the derogatory way in which man can be regarded, by asking:

What is man but a species which, during a brief period, has been dominant over other species on a dwarf planet, revolving round a dwarf sun, which itself is an average undistinguished specimen of a large class of elderly stars which have seen better days ?

When Dean Inge asks this question he is to be understood as asking it rhetorically, with the object of partly answering it.   The question is not intended as an insoluble puzzle ; and yet no one can expect to answer it completely. To say what man is, in his entirety and fullness, would need a knowledge and a depth of feeling far beyond the scope of any individual, or indeed of the race itself. Even such questions as What is electricity ? or What is matter ? are not easy, or indeed possible, to answer completely.

We are constantly confronted with insoluble questions. We cannot completely answer Tennyson's question about a flower in a crannied wall, nor even know all about the pebble which we tread on in the road. A great deal can be said about a flower or a piece of chalk or a lump of coal ; but to dive down into the innermost recesses of reality would involve a comprehensive knowledge of the Universe, which we do not possess. Hence such questions as What is life ? or mind ? or Spirit ? are as unanswerable in their totality as the question, What is the Universe? or What is God ? Yet every now and then we are constrained to ask such questions as What is beauty ? What is inspiration ? What is love ? What is truth ? Volumes have been written round these topics : the aim of philosophers throughout all the centuries has been an attempt to answer them. We cannot fully answer even the question, What is progress ?

Hence when we enter upon a theme such as the subject set down for this course of lectures, SCIENCE AND HUMAN PROGRESS, we recognise that in the very title are terms barely capable of definition—the terms Science, Humanity, and

Progress.   Yet we know that there is a reality underlying each of these terms ; they are not mere words, they have a connotation, some part of which we can understand.   But generalisations and abstract terms are very difficult, and it would be presumptuous to suppose that we can do more than skim the surface, or that we can dive down to any much deeper meaning than has been apprehended by the multitudes who have gone before.

Nevertheless some of us have lived through half of the nineteenth century, and through the feverish activity of such years of the twentieth century as have already sped ; and it is natural that we should occasionally try to take stock of our position, and realise how far we have got, and whither we appear to be tending.   In such topics all are more or less interested when they have leisure to think, and about them the most learned must be liable to make mistakes. Why should we ordinary people attempt to rush in and try to say anything helpful and informing on topics too great for us ?   The deeper problems of philosophy must be for the few ; and yet everyone must have a philosophy of some kind. Even children look out upon their surroundings, and cannot refrain from speculations and conceptions, which they cannot formulate, about the inner meaning of existence, and of all the other strange phenomena among which their lot is cast.

We are living in an age of science, and though the average man knows very little about it he cannot but feel the reaction on himself of the immense amount that is now known. Let us first see if we can attain any view of what science in general is. In itself I suppose it is a system of ordered, systematised, detailed, and as far as possible metrical knowledge, about the things we experience in this world. A system of law and order is found to run through everything ; and a purely scientific man seeks to ascertain and express the behaviour of the common things around him, as in accordance with a scheme of rules or formulæ which he calls "laws of nature." Sufficient progress has been made in this direction to show that in making this attempt we are on the right track. However mysterious phenomena may seem, yet, when scrutinised and examined closely, everything is found to be interrelated with everything else, and everything has an intelligible aspect. We do not know completely, but we know in part ; and there is no reasonable doubt that in pursuing a course of scientific study we are learning something, true as far as it goes, about the processes of nature. The botanist cannot tell us everything about a tree or a plant, but he can tell us a great deal. The physicist cannot tell us everything about electricity and light, but the amount he knows is overwhelming. The physiologist knows an immense amount about the structure and functions of the bodily organs. The psychologist probably knows a good deal about the phenomena of mind ; the naturalist about the instincts and behaviour of animals and birds ; the physician about the nature of disease. Thus science, in order to be comprehended, is split up into departments, which, on account of our human limitations, have to be studied separately and imperfectly, though the philosopher knows that they must ultimately be all interlocked. Thus is constituted the vast bulk of organised scientific knowledge of reality ; and no one knows better than the truly scientific man how incomplete, and in many respects unsatisfactory, his special knowledge is, when regarded from the point of view of the Universe as a whole.

But to the majority of mankind this is not the aspect of science that makes appeal. What they are mainly concerned with is not pure abstract science itself, but the applications and conveniences which they are able to derive from it, and which they apply in the ordinary course of their lives. It is indeed the applications of science that are often thought to conduce to human progress. These applications are not science itself, but are an outcome or efflorescence, a sort of free gift, the harvest or fruits of a deeper knowledge of nature—a knowledge that enables us to control natural forces and utilise them for what we suppose to be our own benefit. And in some cases the benefit is surely real.

But, after all, the power to control the forces of nature and adapt them to our own ends must depend for its value on what those ends are. They may be in the line of progress, or they

may not.   Mere power to utilise the forces
of nature is not everything : and it is not easy,
perhaps it is not possible, to say that the uses we
have made of our present enhanced powers are
such as really conduce to the progress of humanity
in its larger and wider aspect.

To take the most obvious and salient example,
the nineteenth century may be called the age of
machinery.   We have found out how to construct
machines, which shall not exactly supersede, but
shall supplement, human labour to the most
astonishing extent ; and the result, from one point
of view, has been the factory system,—crowd-
ing together a multitude of workers so that they
may be near the machines they have to tend
and by their aid produce commodities more plenti-
fully than ever before.   The factory system has
led to a great output, to wholesale mass-production, and to the cheapening and wide-spreading of
the commodities of life ; but it has also led to
the minute subdivision of labour and to the un-
healthy obscuration of the air.

It is natural to glory over increased powers of
production; and yet we cannot refrain from a
doubt as to their ultimate substantial benefit.
Industrial countries which have devoted themselves to this form of activity have increased in
population and in what seems like wealth ; but it
has been accompanied by great inequalities, by a
great monotony of existence for the multitude, by
a submergence of life under the means of living,
and by a consequent widespread discontent.

I take the following extracts from a paper by
W. H. Warburton—a Birmingham wage-earner
—in *The Hibbert Journal* for April 1926.   He
writes :

> If the workers still remain only partly educated and lazy
> of mind, then the fundamental problems of industry will
> remain untouched. . . . An educated people would find
> means whereby a man's individuality need not be sacrificed to
> machine efficiency.

He goes on :

> The story of the exodus from the countryside to the towns
> is a story of men leaving interesting jobs to become parts in a
> machine that counts fractions of minutes as factors.

And again :

> Workers appear to get used to [uninteresting jobs], and in
> many cases seem to like them.   That is true, but it is also
> true that workers get used to living in slums. . . . This
> most damnable faculty of human beings to adapt themselves to
> any sort of circumstances helps them to survive, but it hinders
> real progress.

Thus it is possible for a pessimist to argue that
the increased power over the material world, the
more extensive use of machinery, the improved
means of locomotion, and the other gains of the
nineteenth century, have had a bad effect on the
human spirit—at least in their present stage of
partial development.   It can be claimed that
corporate happiness was greater in the twelfth
century than in the nineteenth or twentieth, and
that the simplicities of life had after all their
compensations.   It is possible also for critics to

say that men are becoming the slaves of machines, that the artisan is losing his craftsmanship and is in danger of becoming a mere cog in the mechanism. Thereby production is increased; but what is the object of production, and what sort of things are produced ? We are sometimes told that art is decadent, that humanity no longer takes joy in its work, that it is urged by a sort of inexorable necessity to go on producing in order to exist, while existence itself is deprived of much of its value.

These things can be and have been said, and, even though they are exaggerated, there is truth enough in them to constitute a warning. No one can feel that we have reached a stable and satisfactory stage of civilisation. Upheavals and discontent are making themselves more prominent than ever before ; and for this there must be some genuine reason. The discontent is not vicious, not causeless ; it represents the struggle of humanity towards something better. It represents a demand for something higher,—for more opportunity, more leisure, more learning, it may be ; in a word, for more education, for a greater share in the amenities of life, and in the opportunities for developing the human spirit. Production of commodities is not an end in itself ; labour is not an end in itself ; great possessions are not the end and aim of existence. Wealth must surely be that which contributes to well-being. True wealth is that which can be shared, that which makes all richer, and not only a few. Indeed, if

wealth is only understood as inequalities of possession, so that some can live in affluence while
others feel downtrodden or enslaved, such mis-
termed wealth does not really conduce to the
happiness even of those at the top. A kind of
discontent is plainly prevalent in all classes.
Almost proverbially the quest for pleasure,
though enticing, is not satisfying. The real
object is not thus achieved. Self-seeking may
lead to luxury, but it does not lead to happiness.

The deep problems of humanity have not yet
been solved. Many are trying to face them,
and in the effort are doing good work. But the
leaders and social workers themselves are appalled
at the amount of misery and poverty-stricken
existence with which they are confronted ; not
poverty in material things alone, but the poverty
of soul that seems inseparable from the crowded
state of humanity, in mean streets amid sordid
surroundings, with no outlook but daily mechanical labour in factory or mine or office. For it is
surely not the manual labourer only who feels the
monotony : those in the middle classes who
flock into the City each day and spend their
hours in clerical labour must feel it too. Or
if they do not feel it, either for themselves or
for others, and are satisfied, the reason for their
satisfaction may be only that they have begun to
degenerate and are in danger of losing some of
the higher attributes of man.

The nineteenth century was not only the
age of machinery : it was also the age of loco-

motion. Or, rather, the coming of machines has led to a revolution in the methods of locomotion far beyond anything previously possible in the long history of the world. We can go easily from place to place : we have accordingly become restless ; we want to be away somewhere else. The power is good, but is the use we make of it necessarily good ? Surely in the long run it will be. These things are advances ; they lie in the direction of possible progress. But these powers are, after all, only the means to an end, and what that end may be we as yet hardly know.

Not locomotion only, but power of communication has been enhanced beyond all previous knowledge. We can talk now to the ends of the earth. This is a great power, and must surely be of ultimate value. But so far as we have gone at present, though we can speak further, we have no more to say than our fathers had. We are not living in an age of great literature. We can print and disseminate to any extent ; and the result is a mass of periodicals, a quantity of evanescent productions, which have their day or their week and then cease to be. The power of speaking and writing and printing, like the power of locomotion and the power of mass-production, must surely in the long run be good. Every enhanced power over nature must ultimately be a gain to humanity ; it is something achieved that will not be lost ; it gives to the development of the human spirit greater scope, greater freedom, more means of expression. But this utilisation of our gains takes time ; it does not follow automatically. We must not be satisfied with these material achievements ; they are not ends in themselves, but are means to an end : and what the end is, or ought to be, demands from time to time our serious consideration. In my next lecture I hope to consider some of the advantages of these material achievements, and am only now concerned with a warning against glorying in them overmuch.

Although it is necessary to separate the idea of material development and invention from the idea of human progress, and to remember that they are not synonymous terms, and that one does not necessarily involve the other ; yet undoubtedly the one ought to conduce to the other. Every advance in material achievement ought to react, and must inevitably react, on mankind in general ; and every increase in the control over natural forces must be whole-heartedly welcomed. The problem before humanity is to utilise these advances that are put into its hands by the labours of the few, and apply them to the service of the many. This to a great extent is automatically done by the goodwill of mankind in general. For undoubtedly human beings have a good will, and do not really wish to hurt their fellows. Their instincts may largely be trusted ; and material and scientific achievement is all to the good.

But an unreasoning optimism, which sees nothing but good in the advances of the nineteenth century, is not wise, and has been rebuked now and again by the sternness of some prophet such as Carlyle or Ruskin, to whom humanity is too much inclined to turn a deaf ear. Truth is many-sided, and over-emphasis on one side has to be corrected by perhaps over-emphasis on the other. Yet it is an undoubted fact that discoveries may be abused as well as used ; and unfortunately there is not always an international good will. Humanity is not yet united into one family, in spite of the dwindling smallness of the planet. A time will come when that too—the friendly unification of humanity into a family—can be accomplished ; and that great aim is being assisted, rapidly assisted, by the progress of science—though certainly not by science alone.

Meanwhile, and until further advances have been made in that much-desired direction—the uniting of civilised humanity under a sense of mutual obligation and service,—we are living in an epoch of danger. Control over the forces of nature can be applied to destructive as well as to constructive objects. It was a great day in the history of humanity when the first pioneer machine left the ground and rose into the sky. And as that power of air navigation began to increase, you may remember that many people hoped that this new power would never be put to inhuman uses, or, in other words, would never be employed in war. There was a hope that the nations might agree to that limitation. But the temptation was too strong ; a limitation has not been feasible ; and, as Tennyson foresaw, the battles of the future, if there are to be such battles, will mainly be fought in the air :

Heard the heavens fill with shouting, and there rain'd a
    ghastly dew
From the nations' airy navies grappling in the central blue.

Aerial locomotion gives the means of delivering sudden and unexpected blows at a great distance, and overwhelming a whole city in one vast destruction. Every inhabitant will be treated as a potential combatant ; and what we have to dread is a return to the old days of history when a whole population was wiped out, " man and woman, infant and suckling, ox and sheep, camel and ass " ; even the animals being pressed into the service, as they are now, and exterminated wholesale, along with human beings and cathedrals and libraries and works of art.

With every new power, the power of destruction increases too ; and there seems no limit to the destruction that might be accomplished, the damage that might be done, if the whole energy of mankind were directed to that end. Advances in Chemistry are constantly producing new compounds. The power lurking in the atoms, in an apparently quiescent form, is far greater than any of our chemical and mechanical energies : and some of the known compounds are so unstable that their energy can be liberated by a single blow. Hitherto these compounds have

been carried by land and water, and delivered
where it was hoped they could do most damage ;
but now they can be carried far more speedily
by air, and delivered without warning.   Ships
at sea are no longer safe : even hospital ships
have been the object of attack by torpedo, sub-
marine, and mine.   Nothing will soon be lacking
in the possibilities that are available. Methods
of attack proceed more rapidly than methods of
defence : and surely destruction, whether for
attack or defence, is an unworthy object for the
energies of mankind to be directed to, for years
or it may be for centuries.   Surely the efforts of
statesmen in every nation of the world should be
concentrated on an effort to curb and control,
not so much the forces of nature, which in themselves may be beneficent, but what have been
called " the unruly wills and affections of sinful
men."   Patriotism may run riot, as well as other
virtues ; and the nations may vie with each other
in developing their powers of destruction.

The better understanding of and control over
disease may likewise be prostituted to inhuman
ends.   In old days wild animals were used
for purposes of spectacular massacre.   In these
days it is possible to use the more insidious and
still more deadly microbe to the same end.

Indeed, in the present state of civilisation the
power of destruction need not take an active
form.   Mere inertness, on the part of those to
whom the instruments of civilisation are entrusted,
can result in bringing a nation to poverty and

c

misery, without any attack from outside, and without any attempt at active damage. Society as a corporate body has only to refrain from industry, and things will go to rack and ruin around us. It is only by constant labour and attention and good will that the amenities of life are preserved. How soon a neglected building can fall into decay is known to all : machinery can rust and deteriorate : the equalising forces of inorganic nature, which wear down the hills, can soon obliterate the constructions of man. And unless constant labour is put into the land there will be no mines to be ransacked nor fruits of the earth to be garnered.

No scientific advances would be competent to save mankind in such an emergency. The most essential instruments of progress are the old historic human virtues of good will and co-opera-tion. Fortunately this is being more and more realised ; and by the spread of education the danger can be minimised and ultimately overcome. For the mistakes that are now made are not made by viciousness ; they are often the result of ignorance, misinformation, and stupidity. Some-times, indeed, they are the outcome of a self-sacrificing class-loyalty, the antithesis of selfish-ness. Even when the motive initiating a labour dispute is selfish, selfishness is not limited to any one class : the employers of labour in the past have been too apt to regard the manual worker as a mere tool, to be used when wanted, and thrown aside without thought or sense of responsibility.

That state of things undoubtedly has existed, but surely it is already nearing its end. The conscious aim and object of all our activities should be not merely a more rapid production of commodities, but the development of a healthy happy race of beings, who can carry on their work with enjoyment, and develop their lives to the uttermost, on this beautiful planet which is their temporary home.

We are far from that at present. The amount of work that has to be done before this end is achieved is enormous. The era of mean streets and squalid districts in towns cannot be regarded as a satisfactory sign of progress. Looking back even two hundred years, conditions were different ; but we cannot go back, we must go forward. And if we were to come again, say two hundred or it may be two thousand years hence, what should we see ? The result must depend on the labours and wisdom of each generation as it passes over the scene ; but we may surely have hope that in not too long a time the conditions will be almost unrecognisably different. With our present powers things change very quickly. The world is beautiful enough, where man has not spoilt it ; and the activities of mankind should surely improve and develop the resources of nature, instead of spoiling and devastating them.

With improved means of locomotion houses need not be crowded together. Subsistence is not so difficult that the population need be decreased. Food and enjoyment for all are possible, if the labour of mankind were rightly directed, if competition gave way to co-operation, and if each individual truly sought the welfare of the whole. To this end many are striving. The way to it will gradually be perceived, if the effort is made : and the possibilities of life on this planet will be found to be such as have hardly been imagined yet. As a race we are still in the morning of the times. We have hardly begun to tackle the real problems which face humanity. We do not yet realise what life might be : but if each generation strives to leave the world a little better than it found it, there is no end to the good that can ultimately be accomplished.

DESIGN AND PURPOSE

AIMING AT THE DEVELOPMENT OF MAN.

WHEN in the rush and turmoil of life an individual finds time and opportunity for a period of meditation, and tries to realise his place in the Universe and what existence means, there sometimes comes to such an one an overwhelming feeling of Grandeur, accompanied by a sense of personal significance or sometimes of insignificance, which cannot but affect his whole outlook on life thereafter. These times of vivid apprehension may be few and far between : they come sometimes in childhood, sometimes in adult life ; but they are bound to leave their mark, and are experiences not likely to be forgotten. Usually, thereafter, life returns into its old grooves, with an added zest and sense of responsibility : sometimes the whole course of life is changed. Of the former kind a type may be cited in the classical instance of Jacob's vision at Beth-el ; of the latter kind, Saul's experience on the road to Damascus.

But apart from these great upheavals, every individual in the course of his life must surely have some period of insight or of deep questioning. Problems arise to which he may find no answer ; while to a student who has dived

into the profundities of the scientific knowledge
now available to the race, the reality and extent
and wealth and complexity of the Universe
sometimes make appeal with exceptional vivid-
ness. The Prophet and the Poet must surely
thus receive inspirations, which afterwards they
can only try to utter, feeling that words are
incapable of expressing the Reality of which
for a moment they have caught a glimpse. Let
us, in more prosaic mood, seek to set out an
aspect of the material universe as it appears to
a man of science.

MIND AND MATTER, SPACE AND TIME.

First, the Universe is one, and there is no
other. The same system of law and order runs
through it all : the same system of physics and
chemistry holds on the farthest star as on the
earth : the same atoms, with the same rate of
vibration, all obey the same identical laws, and
are all welded together into a visible unity by
the mysterious messenger called Light, which
wings its way across empty space at a barely
imaginable speed, and can anywhere be analysed,
and have its information extracted, by suitable instruments. The message conveyed by
light can be interpreted by the aid of the eye
and the brain, and ultimately by the human
mind, which, though not itself material, has yet
some affinity with the material universe, and is
able to understand and formulate the laws of
its being.

How this interaction between mind and matter came about has been a puzzle to the philosophers of all ages ; but whether we comprehend it or not, the interaction is undoubted. There are some who think that *matter* is dominant, and that mind can ultimately be explained in terms of matter. There are others who take a surely wiser view, who hold that *mind* is dominant, and that all the multifarious phenomena, which can only be recognised by mind, must ultimately be related to, and somehow be expressible in terms of, that mysterious entity. The unification of the two is the ultimate aim ; but the time for that is not yet.

Meanwhile, tentatively and temporarily, we must recognise and attempt to formulate a sort of duality. The universe of matter undoubtedly exists, however we interpret it : while of mind we are directly conscious. We have, therefore, to speak of the material universe on the one hand, of the mental and spiritual universe on the other, in spite of our instinctive feeling that they must ultimately be one. There is, after all, only one Universe, though it may take a great variety of forms. The problem of " the one and the many "—the one ultimate reality with its many modes of manifestation—which was grasped by the ancient philosophers as a problem, is still unsolved. A study of the One is the main business of theology ; a study of the Many is the main business of science. The two points of view sometimes seem incompatible ; but that is only because we do not possess the clue. In moments of insight we feel that there must somehow be a unification, that the whole Universe is a manifestation of the Godhead ; that it really is, as Goethe poetically expressed it, " the living garment of God."

Meanwhile there need be no wonder if the two aspects often seem fundamentally different. It is in no way surprising that controversies have arisen, and that the diversity in our modes of regarding existence has led thinkers from time to time to deny one aspect in order to exalt the other. Ultimate reality, the whole totality of things, is beyond our present mental grasp, and certainly transcends any attempt to formulate it.

Meanwhile, with our senses and our instruments, we explore Reality as best we can, and have thus been enabled to reduce the complexity at least of the material universe to a few elements. Of these I will speak directly, for this discovery is of recent date. But, apart from that, the material universe itself has features that are overwhelming to our finite minds. It is literally boundless ; that is, it has no boundaries, there is nothing beyond it. Reaching out in imagination into the depths of space, we find no limit. Wherever and whenever we imagine ourselves existing, it will always feel like " here " and " now." Everywhere we shall be surrounded by an infinitude of space all round us equally. There is no limitation in one direction any more than in another.

Equally we are immersed in an infinitude of time ; a past without limit, so that there is no beginning ; a limitless future, so that there shall be no end. We cannot imagine a time when nothing existed ; neither can we imagine a time when everything will have ceased to be. Is that an argument for infinite duration ? Yes, I think it is. The human mind is our only weapon of exploration, and its intuitions must sometimes be trusted. Here we are, at the present moment, actually existing. Can we doubt that a sentient being would always have been, and always will be, conscious of that ?

Of any individual world, like this planet, the beginning may be traced, and the end speculated upon ; but of the Universe as a whole that is not possible. Suns may be born, and suns may die, much as a cloud may come into being and disperse ; but the materials of which they are composed existed before, and will exist after. Fundamental things do not spring into being, nor do they come to an end : there is an eternal Now. I am not aware that this has ever been questioned, though some theologians have attributed permanence of existence only to the Deity. Or, as Emily Brontë sang in a moment of inspiration on her death-bed :

> Though earth and man were gone,
> And suns and universes ceased to be,
> And Thou wert left alone,
> Every existence would exist in Thee.

About space, however, the question has been seriously asked : Admitting that it must be boundless, is it really infinite ?—understanding the term "infinity," not in its literal sense of having no limit, a sense well understood by mathematicians, but in the more popular though indefinable sense of actual infinitude. For things can be imagined which have no boundary and yet are not infinite. The most ordinary and crudest analogy, or instance of that kind, is the surface of a sphere. We can move about on the earth without ever encountering a boundary, and yet it is of measurable extent : its area can be expressed in square miles ; it is quite finite, and, being finite, whether large or small makes no important difference. It is the same with a marble or a ball as with the largest star. In that case—the case of a spherical surface—we know quite well what happens : if we proceed continually in one direction we shall find ourselves in time at the place we started from ; we shall have described a contour, without turning round. We shall always have been advancing, we are not stopped by anything, and yet our journey is finite.

Is space boundless in that sense ? We do not know ; but the question has been seriously and reasonably asked. There are evidences which point indistinctly in that direction ; and the attempt has even been made to calculate the length of path we should have to travel before we unwittingly found ourselves returning to the starting-point, by reason of a kind of curvature in space. The idea is not inconceivable, for it has been conceived. If that turns out to be true, the Universe will be self-contained. It has no boundary, there is nothing beyond it, and yet in that sense it would be finite ; its extent could be measured. Enormous of course, illimitable room for everything that exists, but still with a strange kind of limitation. Light, which travels so fast that it could perform the journey round the earth seven times in a single second, may take millions of millions of millions of centuries to encircle the material universe. It is strange that any rational guess can be made on such a topic, strange that astronomers have not shrunk appalled from such a calculation, when even the data on which the calculation is based are unintelligible to the vast majority of educated men. It is not an ascertained truth ; it is a possibility. The progress of science may confirm it, or may discard it. At present we cannot tell : neither can I tell what effect this would have, if it were true, on our general human outlook.

It has been suggested that, in speaking of space and time, we are regarding things too exclusively from the human point of view, and are hampered by our own limitations ; that we have made abstractions not really embedded in the reality of things ; that space and time are an efflorescence of the human mind ; that our sensual limitations to three dimensions of space is an appearance rather than a reality; and that if we could grasp existence, as it really is, we should apprehend things differently and formulate them otherwise. Systems of geometry have been invented in which the mind perceives a consistency quite apart from ordinary experience. It may be that the Universe is infinite in an infinite number of ways, and that our puzzles are due only to our very partial apprehension.

It is likely that there will always be problems far beyond us, however high we may rise in the scale of existence. Meanwhile, all we can do is to explore those that lie nearest us, those that do seem to come within our comprehension, and at the same time keep our minds open to infinite possibilities beyond. Such as it is, the mind is our only guide in these matters; and we shall be wise never to deny the possibility of truths far beyond our conception. Indeed, we have an instinct that, however far we speculate, even though we exert our imagination to the uttermost, the ultimate truth will be better and fuller and greater than anything we have conceived.

OUR COSMIC SURROUNDINGS.

Now let us return to the simpler and more prosaic question of the ingredients of which the material universe is composed, so far as we have yet ascertained. Looking up into space, we see concentrated masses of glowing matter, self-luminous bodies, which have always been the wonder and delight of mankind. At first regarded as mere luminous specks, supplementary to and inferior to the earth, we now know, everyone knows, that they represent an illimitable assembly of other worlds. Of these worlds, some may be dark like the earth ; and those we do not see, except the few that happen to be illuminated by the blaze of the sun, part of the family of which the earth itself is a member, and in many respects akin to the earth. For the most part, all that we see are the self-luminous bodies akin to the sun. We know that the sun is one of the stars. We have learnt something of how they produce their light ; and we surmise that many of them, like the sun, are surrounded by a family of planets, on which, for all we know, there may be life, perhaps not wholly unlike that which we find here. Nothing is known as yet about that ; but the analogy seems too plain to be mistaken. The sun is one of the stars, and, looked at from a great distance, is the only member of the solar system that could be seen. A distant astronomer would know nothing about the earth or any other of the planets, and yet they exist ; and whatever physical cause produced them here may equally well have occurred elsewhere.

We see, then, or it is reasonable to infer, the existence of innumerable habitations, on which life in its various grades may have developed, as it has here. It seems to me absurd to suppose that this earth is the only inhabited world merely because it is the only one we are certain of. What we find here is that life makes its appearance wherever the conditions allow ; it may be vegetable or it may be animal life. Every heap of refuse gets clothed in verdure, and life seems to struggle into existence against adverse circumstances wherever it is possible. We may reasonably suppose that, wherever the conditions allow, the same sort of thing will happen ; and that, amidst the hosts of worlds, there must be multitudes in which the conditions—whatever those conditions may be—are sufficiently satisfied.

GRAVITATION AND COSMOLOGY.

One peculiar circumstance to which we on this planet have become over-accustomed is that every material thing experiences a force directed to the centre of the earth, and that accordingly things cling to its surface with what is known as " weight." Newton found that this property extended to the heavenly bodies too ; that the moon was virtually a detached part of the earth, and was held to it by the same force. A detached body on the earth revolves round it once a day : the weight of a body on the earth is more than sufficient to hold it on. When we lift a body we do not feel its whole weight : there is a small residue which is used in making the body revolve round the earth in twenty-four hours, the amount of this residue depending on the position of the body on the earth, and being greatest at the equator. A body's weight would be sufficient to enable the earth to hold on to it, even if the day were about one-eighth of the length it is now. The day was shorter once, very long ago. A body near the earth, now, if it could fly at five miles a second without friction, need not fly away nor fall ; it could revolve, unsupported, round the earth. The moon is sufficiently detached, so far away, that it suffices for it to revolve round the earth once a month. Again, the earth, which may be regarded as part of the sun, is able to cling to the sun, unsupported, by revolving round it once a year.

Everything is moving ; nothing is stagnant : and in consequence of its locomotion each planet is able to persist. We see the celestial material universe as a sort of law-abiding organism, or organisation, of which the particles are worlds.

There are many modes of regarding the spectacle ; but that which interests us at present is the evidence of Design and Planning which it presents. There is a great variety, but nothing haphazard. It is rather like looking at a work-shop of running machines : they appear self-working and complete, but they are full of evidence of design to accomplish a certain object. From what we know of the details here, it is legitimate to surmise that the object

of all the worlds in space is to provide opportunity for life and mind to develop, and that
the arrangement of the matter in the Universe
is subsidiary to that great end.   The stars, for
instance, whatever may have been their initial
stages (through which they pass fairly rapidly),
are of a size well adapted to supply heat
and light and other necessaries to any planets
which surround them.   If they were not so
big they would not be hot ; or at least they
would not be able continually to evolve the heat
which they radiate : they have stores of energy
well adapted to their purpose.   Yet, however
strongly we are impressed with design, the
object aimed at is always accomplished through
a rational physical cause.   We must be ready to
look for the physical cause, and never be perturbed by finding out the process and the way
things are arranged.   Too often has there been
this perturbation : people will not realise that
the Deity acts through agents and by physical
means.   Consider the stars again ; we might have
imagined that under their mutual gravitation
stars would, in the course of time, aggregate
together into one great mass.   The distribution
and separation of them throughout space is
plainly advantageous : it must be due not only
to intention, but also to some physical cause.
That cause has been expounded by our great
astronomers, who have shown that if any star
were immensely more massive than the sun it
would not be stable, but would break up.   Many

such masses have broken up, and the result is the separation and individualisation of matter distributed throughout space. Eddington has shown that the internal constitution of those stars which radiate freely is a nice balance of energies, equally shared between matter and ether ; and that this determines a rough uniformity of mass through all the brighter stars.

Can anyone rationally suppose that all this is accidental ? To the eye of the imagination, difficult as it may be to reason it out, the evidence of some sort of Design and Purpose is conspicuous throughout the Universe as a whole, and is as clear on the large scale, as in smallest details. Nothing is too great, nothing too small for attention. Everything that exists appeals to the human mind and indicates a kinship with its rational processes. Our senses enable us directly to perceive only the superficial aspect of the universe of matter ; the rest is all inference, but legitimate inference. Speculation and intuition outrun reason, but sooner or later they come within reason's scope : and the more we probe, the more we find that, in spite of outstanding puzzles and problems at present insoluble, Wisdom is justified of her children.

Our Material Basis.

We have long known that matter consists of atoms, atoms of ninety-two different kinds, which seemed at one time to form the ultimate substratum of which everything was composed.

We now know that these units are not ultimate, but that, like solar systems, they can be analysed into still more fundamental ingredients ; and those ingredients we have recently ascertained to be nothing more or less than what we had previously observed, imperfectly and partially, as positive and negative electric charges.

Here again there is a sort of dualism, which some day will be resolved or unified. At present we do not know the ultimate nature of an electric charge ; but we do know that there are two different kinds, one more than a thousand times as massive as the other. Both have the fundamental property of matter, namely, inertia. Both are subject to violent forces, which we call electric forces ; both when moving exhibit the phenomenon of magnetism; and, when suddenly stopped, both emit some kind of radiation, the best-known kind of which is light. Last century we thought we knew something of the nature of light ; now we are in doubt. Neither electricity nor magnetism nor light is fully understood ; and yet the laws of their interaction have been elaborated in a remarkably complete manner.

The opposite electric charges, though they attract one another violently with a force much greater than gravitation, are yet kept separate by their motions. They weld themselves into systems and patterns of singular intricacy and beauty, each unit revolving in a regular orbit with as much accuracy and precision as any of the planets. Law and order reign supreme not only in the heavens but in the interstices of the atom, which, after all, is built on much the same pattern ; so that, looked at with the eye of science, any piece of common matter, such as stone or wood, resolves itself into something like the Milky Way—a collocation of separate particles, each of which, by obeying the laws of its nature, subserves the functions of the whole. The electric units build up into atoms ; the atoms aggregate into molecules ; and, strange to say, when these molecules attain a certain complexity, they may become animated and display the surprising phenomenon of life.

Life and Matter.

There are those who think that life is a mere consequence of complexity of structure. There are others who think of it as the rudiment of something immaterial—more akin to the mental and spiritual aspects of the Universe—which for purposes of manifestation and development and investigation can interact with matter ; interact, that is to say, with those molecules of matter which have a structure sufficiently complex to be in that way utilised. Biologists tell us that the inorganic passes insensibly into the organic, changing first into something allied to the vegetable, under the influence of the rays of the sun ; and that it is by the aid of these vegetable products that animal life has become possible. Animals cannot transform the inorganic ; only vegetation does that : but on the plant structure, thus provided, animals can feed and derive their sustenance ; while their products, in turn, assist vegetation to assume still more complex and beautiful shapes. Thus the whole of matter— animal, vegetable, and mineral—is welded together into a coherent and co-operative harmony, each being necessary to the others.

The noteworthy thing throughout Nature is the brilliance of the result achieved compared with the simplicity of the means employed. In the material universe we now know there is nothing but groupings of the two electric charges. Enormous in number, they form the ultimate units of which everything is built. Apart from those units, no matter would exist ; but with them, under the laws of electric forces and the superadded force of gravitation, the whole magnificence of the material universe has been constructed. It is an amazing outcome to follow from the utilisation of two fundamental elements. The lavishness of Nature is prodigious, and is itself a kind of infinity.

Whether the two electric units are really fundamental, or whether they in their turn can be analysed and resolved into something still simpler, we cannot at present say. For myself, I can hardly doubt that they are capable of resolution. Indeed, I hope that before long that further step may be taken ; but the time is not yet ripe.

Meanwhile there is plenty to contemplate in the analysis so far made ; and anyone who succeeds in realising it to the full must be over-whelmed by the contemplation. Truly the Universe does seem infinite in an infinite number of ways.

All this may seem far removed from daily life and the ordinary commonplaces of human effort. But we are a part of the Universe; and an apprehension of its nature, even on the physical side, cannot but have an effect on our outlook, and may help to raise us above the changes and chances of this mortal life. Such ideas cannot always be before us, but in the background of our minds they exist, and must have their due influence. They help us to take a wider and broader and truer view of human destiny. They enable us to see that there is nothing small or insignificant, and that everything is arranged for some great and far-seen Purpose, towards the furtherance of which each in his day and generation is privileged to help. Life becomes a nobler and grander thing ; and even the sacrifice of life for the achievement of some end can be felt to be a reasonable and unwasted contribution. There is lavishness in Nature, but no waste.

At times it is difficult to realise this. Now and again in our short life, or apparently short life, and with our very limited outlook, we may be submerged in distress and helplessness and anxiety. Pain may confuse our faculties ; bereavement may cast its shadow ; but from those to whom Vision has been vouchsafed, the light

is never completely withdrawn. They know
that all this extraordinary manifestation of Design
and Purpose must have some unknown object,
that there must be some far-off Divine event to
which the whole creation moves ; and those to
whom the revelation has come take comfort in
the thought that underneath are the Everlasting
Arms.

Some Proximate Steps towards Progress.

In more prosaic mood, let us now review both
the aids and the hindrances to human progress
resulting from the application of science, so far,
and more especially the contributions to man's
development which might be made in a reasonable time. A full and complete application of
scientific knowledge must ultimately contribute
to health and enjoyment ; but meanwhile science
half applied has crowded people together into
tenements, and destroyed the beauty of the
country for miles around. It has also, most
unfortunately, flooded the atmosphere with smoke,
and excluded the health-giving rays of the sun.
Not, indeed, that that smoke cuts off the heat
and light to any great extent, save in extreme
cases, but the thinnest coal-smoke impurity
quenches the ultra-violet or short waves which
have such a beneficent effect on all the processes
of life. Those short waves, or actinic rays,
thus cut off, would lessen the tendency to
disease, and in some cases actually effect cure.
Until the towns are improved, that is, until the

places where human beings live in large numbers
are made healthy again—by improved methods
of utilising energy without the waste and ugliness
which at present accompany that process—the
only way for the present generation is to get
away from towns occasionally ; and the facility
of locomotion renders that possible.

It is not by any means a final or ultimate
solution ; there ought to be no need to search
for healthy air ; but still, movement in itself is
pleasant, and experience can thereby be enlarged.
Whereas our forefathers were limited for the
most part to their own districts, now it is possible
to visit far-distant countries, and thus establish
friendly intercourse with people whose very
existence could not previously have been effectively realised. The more we know of mankind
the more friendly we feel towards it ; though
certainly it behoves those who travel to regard
themselves as to some extent responsible for the
idea of their countrymen which the inhabitants
of the visited countries will form, and although
at present there is often too much selfishness
and thoughtlessness in the behaviour of people
on the open road and in the foreign countries
that they visit. Ill-bred picnic parties strew
paper, but ill-bred travellers strew ill will. The
increased facilities which science puts into our
hands should not be utilised thoughtlessly. A
sense of responsibility should accompany their
use ; and those who travel may well feel that
even their small efforts can help towards a more

friendly feeling among the nations, and so gradually lead to a kind of understanding and sympathy that will make wars impossible. And now that speech is possible, or soon will be, all over the earth, nothing but the troublesome difficulty of language prevents the mutual understanding which must ultimately follow. If all the nations could co-operate, and each contribute its quota according to its special facilities, with free interchange of products, what an advance that would be !

Science has made this possible on a large scale ; the ends of the earth are brought nearer together. Fruits can be garnered and transmitted and consumed without such lapse of time as would lead to their decay. In some directions, therefore, the free interchange is already beginning. Especially is this the case with scientific discovery : every advance made in any part of the world is communicated and utilised by the whole. There is no monopoly in such things, beyond that which is necessary to initial development. All knowledge, all art, everything of real and permanent value, is free and open, and becomes more valuable the more it is shared.

That is indeed, as I briefly hinted before, the test and criterion of true wealth. The sharing of true wealth does not diminish our own possession, but increases it, and increases our joy in it. The more we can get people to realise the beauties of the earth, the higher and

keener becomes our own appreciation ; and it
is surely to this end that artists of every kind
—musicians, poets, painters, sculptors—are working. They try to re-create things so that all
can enjoy them. There is no selfishness in
true wealth ; and there is probably no joy
comparable to the privilege of being allowed to
contribute to the well-being and welfare and
advance of humanity.

Surely it is for this that statesmen are labour-
ing ; it is for this that power is valued. Many
a rich man has realised that the selfish adminis-
tration of his property gives no satisfaction, and
he bestirs himself to think how he can best
and most wisely dispense it, so as to do least
harm and greatest good. I do not say the
problem is easy ; if it were easy it would not
be a problem ; but more and more are the
employers of labour realising that the welfare of
humanity is their ultimate end and aim, and the
only one that can be really satisfying ; they are
more and more trying to organise their factories
so that the workers, too, may get some joy in
their work and share the benefits of the rapid
means of production which science has put into
their hands.

Undoubtedly there is much to be done in
this direction still ; lamentable disputes between
capital and labour are far too prevalent. We
are living through a troublous time, when the
solution has not been found ; but many are
seeking it ; and some are now beginning to

realise, and to say clearly enough, that a solution will not be found without some mode of profit-sharing between all those who are co-operating in the work. There is far less discontent when the workers feel that they have a corporate interest in it and are able to know what is going on. If there is a Board of Directors, they should have a representative on that Board ; they should realise the difficulties of management. Occasionally such a selected representative of labour may be able to make no little contribution towards the enterprise, the forethought, and the planning, necessary for any great undertaking. Depend upon it, a large amount of ability is lost for lack of opportunity to apply or use it. In exceptional cases a lowly-born genius, endowed with strong character, presses upward in spite of obstacles : we know that it is possible ; but it must be exceptional, and many must fail. Yet ability is limited to no one grade of society. Among the workers there must be a lot of ability lying latent and undeveloped : much of this might be pressed into the service of industry by a suitable partnership system.

Those who have associated with artisans, with laboratory assistants, with many in the intermediate grades of society, have been impressed with the ability which they find there, and have been inclined to lament the lack of opportunity for its free and complete development. A Faraday occasionally emerges, and by his quiet

laboratory work effects discoveries which lead
to the initiation of a whole new profession, like
electrical engineering : and among the workers
there may be now, and in all probability are,
some potential Faradays. The workers are
beginning to seek opportunities for education.
They are rather too apt to take a materialistic
view of it ; they hardly know exactly what they
want ; but they are not content with their
present outlook, nor should they be. Reorgani-
sation is required : the difficulties of the present
situation will compel it. No outsider can lay
down with wisdom what that organisation ought
to be. It must come, presumably, like other
things, through a process of evolution ; but it
can be aided by conscious effort, and by the
growing sense of responsibility and good will.
Another century will see a great change in the
conditions of labour, and in the relations between
labour and capital ; and when the nascent ability
of all mankind is liberated, it will seem astonishing that we have been so slow in recognising
the power that we all the time possessed.
The fruits of truly co-operative industry, and the
widespread feeling of good will and common
interests among all concerned in the activities
of life, will conduce, perhaps more than anything else that we see within our grasp at
present, towards the true and lasting progress
of mankind.

For, depend upon it, if internecine class
warfare were abolished, and if humanity were

liberated from the restricting monotonous conditions under which much of its work is now done, the gain would not be in material development alone.   There would be more leisure, more happiness, more joy ; the soul of man would be liberated for higher things ; and the spirit of mankind in general might rise—as in time it will and must rise—to the spiritual heights attained at present only by a few.   Contrast the mind of an Eddington with the mind of a typical Russian peasant : it seems marvellous that they belong to the same species !   Yet the potentiality is there.   What man has done, man may do.   Under better conditions the human race may rise to heights unspeakable.   We little know what in a million years man is going to be.   I am speaking of the race—man on this planet—for science tells us that the planet and its activating sun can last far longer than a million years, perhaps a hundred million ; and in that lapse of time almost anything is possible. As Browning truly says :

> . . man is not Man as yet.
> Nor shall I deem his object served, his end
> Attained, his genuine strength put fairly forth,
> While only here and there a star dispels
> The darkness, here and there a towering mind
> O'erlooks its prostrate fellows : when the host
> Is out at once to the despair of night,
> When all mankind alike is perfected,
> Equal in full-blown powers—then, not till then,
> I say, begins man's general infancy.
>   —BROWNING : *Paracelsus.*

Surely such an application of science as radio-telegraphy is an unmitigated blessing. Radio-telegraphy was foreshadowed dimly less than forty years ago ; within that period the very first discoveries were made which rendered it possible. Not science alone, but the applications of science, have borne fruit prodigiously during the last half-century. And we need not think of physical science alone : biological and medical science have advanced by leaps and bounds. Steps towards the prevention of disease are being taken all over the world. The life of humanity is being lengthened ; tropical countries are being made healthy ; the opportunities of humanity are being increased ; even the powers of humanity are enhanced beyond previous conception.

And in another region, I venture to say—a region not yet recognised by orthodox science—evidence is accumulating that humanity as a whole is not isolated in the Universe, as it used to think it was, but that we are in close and affectionate touch with a higher order of beings, who realise our difficulties, help our struggles ; and who, recognising the vital importance of this earthly period of existence, are straining their faculties to the uttermost to step in wherever they are given an opportunity—not by force, not with any compulsion, but by permission, by good will, or in response to entreaty—so that, by co-operating with us, they can contribute to the advancement of the whole.

The Christian belief is that not only are there higher powers thus operating, but that some so lofty that we can only attribute to them divine attributes, and speak of them by some name appropriate to the Deity, are sharing in the struggle and effort, have been willing to undergo sacrifice, have even entered into the turmoil, if by any means the race of struggling creatures now toiling blindfold on this planet, little knowing of their importance and their destiny but each striving instinctively towards some unknown end, may be assisted and helped out of the mire, their feet set on firm ground, and they themselves encouraged to march face foremost towards the Coming of the Kingdom of joy and faith and peace.

HELP AND GUIDANCE

I SPOKE last time of the evidence of design in the
Universe and its implications as regards humanity. Planning and purpose cannot be excluded
from the Universe, for we are conscious ourselves
of possessing and exercising those powers; and
whatever we have is thereby proved to exist
in the Universe, for we are in it : the only
question is their scope. Design and purpose and
planning, if admitted in the Universe, seem to
carry with them the idea of possible help and
guidance going on now—that is to say, creative
and productive activity not only in the past, but
in the present too—continual interaction with the
material and even occasional interference with
the physical order ; in much the same way
perhaps as we and other live things interfere
with it. We ourselves are not limited to the
region of physics and chemistry : our minds are
realities too, and by taking thought we can affect
occurrences and bring about planned results in
the material world. The Universe is far more
than an aggregation of matter, however beautifully organised and adapted to its purpose that
matter may be. There is an unseen and unsensed

universe too, a universe of life and mind and spirit, which clearly dominates the material, and which, though it makes no direct appeal to the senses, is equally real.   I shall soon deal with the reality of this unseen or unsensed portion ; but first let us consider some of the difficulties which from time to time have been felt on the physical side by men of science, and caused them to deny the possibility of help and guidance.   Some have tried to look at mind as if it were a sort of efflorescence or outcome of material organisation, which did not really affect anything, and was more apparent than real.   From which it would follow that the interfering activity of higher powers was still more superstitious and absurd.

DIFFICULTIES ABOUT INTERACTION.

In speaking of the interaction of a spiritual with the material world, difficulties have often been adduced from the scientific point of view— difficulties caused by the conservation of energy and the supposed completeness of our knowledge of physical laws, so that interaction and interference cannot really occur.   The objection was urged by Tyndall with special reference to answers to prayer.   He urged that interference with the course of completely settled and law-abiding nature was impossible.   But the same objection would apply, if it were valid, to guidance generally.   On that basis we could not deflect a river or plant a tree or pump water uphill ;

and it would be presumptuous to ask a gardener to water the garden, for it would be attempting to produce a local and temporary shower which was not prearranged. If the course of events is inexorably fixed (as in a purely materialistic view it presumably is fixed by the interactions between atoms and their available energies), then, it has been argued, unless life and mind were themselves forms of energy, the process would be complete without them, and any interference on their part would be out of the question.

Such a conclusion, however, is clean contrary to experience : we all know that live things can produce results that would not otherwise have occurred, from a bird's nest to a palace, from the song of a lark to an oratorio. We know that they do this by utilising the solar and other energies available ; they do not increase or create any energy, but they guide and direct the available energy into unwonted channels. That is, indeed, what life is, a guiding and directing principle. It is powerless to act unless energy is supplied ; but it can determine the result of the working of that energy : mind is able consciously to plan and determine how energy shall be utilized. Thus it is that rails are laid down to guide a train to its destination. A motor-car needs constant attention on the part of the driver : on a railway the direction is determined beforehand, and all that the machinery has to do is to propel the vehicles ; but the propulsion is effected in accordance with a previously planned time-table. Life and mind are everywhere dominant, especially in timing and aiming. The powder only propels the bullet; it does not aim: nor can machinery discriminate between friend and foe.

Still there is a legitimate question, How can mind, or indeed life either, both of which are apparently outside the scheme of physics, enter into that scheme; how can they direct and aim and time the movements of matter? Everything has a physical concomitant: what is the physical concomitant of guidance? A little time ago we should have answered that a guiding influence was exerted by a force at right angles to the line of motion, one that did no work, consumed no energy, but merely curved and changed the path. But now the whole idea of "force" is under a cloud. True, it is a direct experience; we know when we exert force by means of our muscles; and we can see similar effects produced in other but apparently to some extent similar ways. Hence the idea of force is in no way a recondite one; it is a thing of every day; but yet it is not easy to explain: an attempt at an explanation would be elaborate, and even then unsatisfactory. We must take refuge in direct experience and learn from facts. How we move our finger or our hand, say in writing, we may not fully know; but the fact that we can do so is obvious. In every case the balance-sheet of energy is complete; energy put in at one end appears at the other end of the process: there is no increase, and there is no loss. Everything physical about the process of writing or speaking, or fighting or climbing, or any other action, is subject to the laws of physics.

Life and mind do not abrogate or in the smallest degree falsify those laws, but supplement them ; so that the result is not merely mechanical movement, but sense. The sense of a message is outside the scope of physics—a meaningless phrase may require just as much energy to utter as a sane one. An order to a troop may change the course of history ; but it does so in no mechanical or materialistic way. The recognition of friend or foe is not a question of physics : the optical impression, apart from mental activity, may be just the same. It has been pointed out that the material form of a message is quite irrelevant ; it may be spoken in any known language ; it may be written on a piece of paper ; or it may be telegraphed. The emotion excited by such a phrase as "Your brother is shot dead" does not in the least depend on the material vehicle through which it has come, nor on the language drawn upon for words. *Dein Bruder ist erschossen*, or *Votre frère est fusillé* does just as well. The sound form is different, the emotion is the same. The emotion is called out by the idea, not by the physical manifestation or incarnation of that idea. The idea received is in accordance with the intention of the sender of the message, who then directed some available energy to a foreseen and planned end.

We all know, therefore, that guidance is possible ; and any scepticism about guidance from a spiritual world must depend, not on physical difficulties about guidance, but upon doubt whether intelligent beings with guiding powers exist. Once their existence is admitted (and that clearly is a matter of evidence) there remains no further difficulty about the admission of their possible activity in influencing mankind and regulating affairs. They may or may not do it ; that again is a question of fact ; but there need be no question of physical possibility. We learn this by analogy with what our own powers are. We know by actual experience that intelligent beings like ourselves can plan and help and interfere, and can produce results that without us would not have occurred. The whole of Architecture and Engineering and all branches of Agriculture depend on that power.

A Plea for Simplicity.

We as higher beings are constantly influencing the behaviour and actions of the lower creatures : we can reward and punish them ; we can indicate affection or dislike, and they are able to indicate appreciation of our attitude ; we can entrap or set them free. We can, in fact, perform actions which to them are miracles. We often interfere in their conditions without their knowledge : we give them opportunities of service, we appreciate their help. In many and various ways we guide their destiny. It is absurd to say that a higher being cannot control or influence a lower. The lower animals are well aware of some of our powers, and they sometimes petition us to use them : if a cat wants a door opened, it asks us to open it. If animals are in trouble or distress they cry out ; if they are comfortable they can indicate pleasure and gratitude. Some creatures, indeed, are quite unconscious of our activities, probably unaware of our presence ; but we act on them just the same. To say that prayer to a higher being is inoperative and cannot be answered is absurd. There is no justification for such a view in connexion with the lower animals ; and the only justification for it in connexion with humanity is the gratuitous assumption that no higher being exists.

If anyone is able to contemplate the Universe in all its magnificence and interlocked beauty and variety, and come to the conclusion that nothing higher than mankind exists in it, I cannot envy him his common sense. The Universe is shoutingly full of design, plan, intention, purpose, reason, and what has been called Logos. Without it was not anything made that was made. Not only the heavens, but the earth ; not only the flowers, mountains, sunsets, but every pebble, every grain of dust, the beautiful structure of every atom, proclaim the glory of the Being Who planned and understands it all.

Shall a conviction of this kind have no effect on the progress of humanity ? Surely the true progress of mankind must depend on its realisation. The race of man must learn that help is forthcoming if it is asked for ; that though guidance may be acting in ways we do not know, though there is much of which we are unconscious, yet petitions are not inoperative ; the will to receive makes it possible to grant ; the desire to be guided makes guidance easier. The common-sense aspect of these things, which children easily apprehend, an aspect free from the subtleties and difficulties which further study may inflict upon them, is after all a truer mode of representation than our fine-spun theories can achieve. We may easily lose the way in man-devised schemes ; but the great fundamental things are simple ; their perception is derived from very commonplace experience. The path towards truth in such matters is plain and straightforward ; and the wayfaring man, even though a fool, need not err therein.

My message, therefore, in these matters is one of great simplicity—a plea for return to a childlike attitude. Whatever we want, we can ask for. We may not get it ; we should not really wish to get it unless it is wise ; we are not putting forward our own wills, but only our longings. It may be they are legitimate ; it may be they are not. We cannot hope to understand ; and if a petition is not granted we need not rebel. All things are in higher hands than ours. "Our wills are ours to make them Thine." We are a part of the whole, but how small a part, and how little we understand !

Yet we are not wholly ignorant : we can learn from analogy, learn from the beasts and birds around us, learn from every activity in Nature. And thus arises a kind of natural and simple theology, a recognition of the goodness and love which underlie all the complexities of existence, the perception that even pain and evil have a meaning which can some day be recognised, and that the whole of existence is working together towards some far-off Divine Event, without haste, without cessation, one stage after another being gradually attained.   And all that has been attained is seen to be good, even though in the infinitude of time it is but a step to something still better.

Trustworthiness of the Senses.

Humanity learns about itself and the Universe through its senses. Our primary senses, I suppose, are those of heat and cold, pleasure or pain, also hunger and our other appetites of a periodic nature ; but in addition to those self-concerned sensations we possess extraordinarily ingenious specialised sense organs, whereby we apprehend something external to ourselves. Taste and smell are of an internal kind, rather like the sense of temperature or of pain ; but they are interpretable also as the results of causes independent of and outside our own bodies. A smell may suggest a flower, or it may lead us to examine drains. Similarly even a pain may tell us something ; the pain of a sword-thrust differs from the pain of a headache, and rouses quite different emotions. We have thereby learnt about something that occurred in the external world.

By our muscular sense we appreciate force, the resistance of solids in the external world, the general inertia and impenetrability of matter, with all manner of gradations and fine shades which we summarise under the sense of touch. Then again we appreciate in the most lively way certain vibrations in the air, and thus construct an external world of sound and music. We have a sense organ, also, capable of being affected by far more rapid vibrations in the ether. These we interpret, not as vibrations at all, but as a representation of outside objects which have in some unknown manner influenced the ether of space. Ordinarily we know nothing of all this, however ; all that we really apprehend is what we call a " sight " of the object—that is, a direct mental impression ; all the rest is the result of investigation and discovery. There is no time to elaborate the structure and design of the ear and the eye. Suffice it to say that they are splendid instruments, that to them in combination with the muscular sense we owe the substratum of all our information. They give us true information as far as it goes ; they do not deceive us unless we maltreat them, and even then they give true indications, a true report about that which is really presented to them.

It is sometimes said that the senses are deceptive ; but, like all other physical instruments, the deception is not in them, but in our interpretation, our reading of their indications. To take a trivial example : If iron and wool at the same temperature are touched, the iron feels cold and the wool feels warm. Why ? Because the part touched really is cold in the one case and warm in the other : the hand is chilled by the iron and not by the wool. The nerves embedded in the skin can only tell the temperature of the skin, and that they tell truly. The iron chills them by conducting the heat away, when the hand tries to warm it ; so the iron feels cold because it really is colder than the hand. But any heat imparted to wool stays where put, and accordingly the part touched feels warm because it really is rather warm. It has been warmed by the touch. The heat was not dispersed or carried away.

This instance can be generalised. Every physical instrument indicates its own condition, and when moved into new surroundings it may modify those surroundings. It may be influenced also by the presence of an observer ; it will not tell us the condition that was, any more than the condition that will be ; it tells us the condition that is. Remembering all that, we can bestow unlimited praise on our sense-organs, and can learn to interpret their indications correctly.

TRANSCENDENTAL INTERPRETATION OF SENSE PERCEPTION.

But now let me remind you how little our senses, even our highest sense organs, tell us

directly about the Universe; how little they
tell us about any object in it.   The reality of
things lies not in what we see or hear or feel,
but in what we infer.   Our senses we share with
the animals ; indeed, we derive them through our
animal ancestry.   Some animals can see and
hear just as well as we can, or even better ; but
how little they can understand ! The difference
between us lies not in our senses, but in our
minds : we have a power of inference denied to
them ; and it is through the mind that we
apprehend reality.

Take a few obvious instances.   The essence
of a work of art, say a painting, lies not in the
pigments and canvas which we see, but in the
intention, the meaning, the soul of the artist.
On its physical side the greatest work of art is
insignificant : it is either a block of marble
carved to a certain shape, or it is a cunning
arrangement of coloured earths on canvas, or it
may be a succession of harmonious sounds, or,
again, black marks on a piece of paper.   A
poem's real existence is not on paper, but in the
mind.   It was conceived first in the mind of the
poet ; it was then recorded mechanically in
accordance with the convention of language ;
and thereafter was left for a sympathetic mind to
discern and understand.   The conventional symbols have the power of arousing, in a receptive
percipient, ideas akin to those of the poet ; and
apart from that perception and interpretation the
poem can hardly be said to exist.   To a savage or

an animal it has no meaning : even to a civilised man it has no meaning unless he knows the code, or can get it translated. The poem exists not in the region of words and symbols, but in the realm of ideas : its reality is purely mental. An animal can see the paint and canvas, but to it there is no picture : an animal can hear the noise of a symphony, but not Beethoven.

What we appreciate in a work of art depends on what we bring to it. If we have no feelings akin to those of the artist, his work is lost upon us. Physically considered, a cathedral is an arrangement of stones and wood ; while the tones of an organ are but vibrations of the air. Mere sense perception, like all ordinary everyday knowledge, is very superficial.

> For Knowledge is the swallow on the lake
> That sees and stirs the surface-shadow there
> But never yet hath dipt into the abysm,
>
> .   .   .   .   .
>
> And when thou sendest thy free soul thro' heaven,
>
> .   .   .   .   .
>
> Thou seest the Nameless of the hundred names.
>
> .   .   .   .   .
>
> Thou canst not prove the Nameless, O my son.

It is instructive to consider the method by which we communicate with each other, the strangely indirect process that constitutes speech. Somehow, mysteriously, an idea conceived in the mind, whatever that may mean, is able to operate on the cells of the brain, and causes them to transmit an impulse down the nerves till they reach the muscles. The muscles contract ; the slit membrane that we call the larynx is partly closed and tightened up ; the resonant cavity above is adjusted ; the lungs contract and expel air through the orifice, and thus, as in any other musical instrument, the air is thrown into vibration : a kind of vibration thoroughly understood and analysed, consisting of a series of compressions and rarefactions following each other in regular order. These pulses spread out in all directions : the greater part of them fall upon inert material, are quenched, and turned into heat ; though they may rebound several times first. Some small portion of them falls upon a stretched membrane or diaphragm in our head, which by a system of levers transmits them to a fluid in which nerves are embedded. These nerves are associated with a complex stringed instrument or membrane, each string or fibre of which is able to respond to some one tone or rate of vibration. The nerve-endings so stimulated are collected into a bundle of fibres called the auditory nerve, and thus transmit, not indeed the original pulses, but presumably some form of energy, to the appropriate brain-cells.

So far, with some lacunæ, we are able to follow the process fairly well. It is a physical and physiological animal process, such as many sentient quadrupeds could share. But now again comes a region of mystery : the brain-cells are in touch with the mind. " Touch " cannot be

the right word : the right word is not known.
But somehow they can affect the mind, and
there reproduce the idea or sentiment or
thought or emotion responsible for starting the
whole series. One mind has communicated with
another ; the other mind understands, and can
make response by a similar indirect and round-
about process. Response may take many forms ;
any muscles in the body may be actuated. The
response need not be a vocal response ; it might
take the form of something done, a nod or a
frown, or an act of approach or recession. The
feeling aroused might be neutral, or the result
might be joy or grief or anything correspond-
ing to the idea conveyed.

Speech is not the only method of intercourse.
We have learnt how to communicate in writing,
even with people at a distance ; and still more
recently we have learnt how to communicate
telegraphically, utilising the ether of space for the
purpose, by means of certain instruments invented
as supplementary to our physiological organs. In
these ways we transcend the powers of animals
and of uncivilised man ; but still the process is
indirect, the physical part of it contains no intelligence, no emotion, nothing of the real message :
it can only serve as a detent or trigger capable of
calling out the idea in a mind capable of realising it.

It is amazing what various forms a message
can take in its intermediate stages. It is the
same with a mathematical calculation : an idea
or some data are put into symbols at one end,

the symbols are juggled with or transmuted in accordance with certain rules, and an enlarged or modified idea emerges and is interpretable at the other end. In that case we have to attend to the intermediate steps, though only to see that they are performed correctly ; but in ordinary speech, to which we have grown so accustomed that we fail to realise its wonder, the intermediate stages work automatically ; we do not attend to them at all : most people forget that they exist, or never knew.

Similarly when engaged in expounding the method of vision—how the atoms at one end generated or modified ether waves, which then travelled at an immense velocity, and were focused by the lens of the eye, in accordance with the laws of optics, upon the retina ; how there, in some only partially known way, they stimulated the nerves, which conveyed their stimulus to the brain, and so on to the mind, which was able to form a clear image of the object which had modified or generated the waves ;—I have known people say, after all that had been explained, "But what need for all that elaboration ? What I look at I simply see ! "

Yes, that is just a parable of how we regard the Universe in general : we simply see it, or we hear it, or we feel it ; and there seems nothing more to be said. In fact, there is everything to be said ; and generations of workers may pass before we fully understand the mystery of sight. It is more mysterious than sound ; for there the energy resided wholly in matter ; but in vision, the energy, though it started in matter, is transmuted into a disturbance of the almost unknown connecting medium, the ether, and then has to be transmuted back again. The two processes of conversion are called radiation and absorption, —words easy to say, but about which volumes have been and will be written.

It is most difficult to understand clearly, indeed we still only partially understand, how the energy gets out of atoms into ether, and how—after, it may be, a journey of thousands of miles, or, in the case of the stars, millions of millions—it gets back again, so as to be capable of interpretation. Yet so it is ; the human mind is capable of interpreting the messages of the stars. The light which comes from them is full of information. No, that is hardly a right mode of expression. There is no information in the light : there is nothing but a peculiar tremor in the connecting medium. The information, the reality, is entirely in the mind.

Through all the generations of mankind, till now, the message was unread and uninterpreted : though the indications of sense were the same then as now. Immense ingenuity and learning, such as the average man knows nothing of, have been expended in the quest. Highly ingenious instruments are necessary ; and even those tell us nothing directly. They may sort the light out into patterns : they may concentrate it in certain positions, and remove it from others : they can form what is called a bright-line spectrum, the position of each bright line being accurately measurable, with the intervening spaces dark. But what of that ? Who can tell what all that means ? Who can read the map and tell from it, not only the chemical ingredients of the distant star, and how fast it is moving to or from us, but also its age, its constitution, its history, the family to which it belongs—any amount of real knowledge ? Who but those whose minds have been cultivated, and whose studies have been specially directed for years, to enable them to understand and interpret ! Even the actual size or diameter in miles, and the average density of material of which the most distant bodies are made, can be ascertained by the message, the inorganic and merely tremulous message which the ether brings. The ripples in a harbour are modified by boats and breakwaters and other obstacles, and may be said to indicate the nature of these obstacles by the pattern of the ripples. But who could hope to infer the objects from an examination of the rippling surface ? Yet that is what we do every time we open our eyes, or when we snap the shutter of a camera on a landscape : every detail is depicted on the sensitive surface, shape and colour, all complete. In the eye, the image vanishes almost directly we close the orifice : in the camera, the light and shade can be permanently recorded by a chemical surface specially contrived for the reception of the image.

Yes, the process is intelligible up to a point ; there is not much mystery about a photographic camera, though there is much food for thought and much possibility of physical explanation at every stage. The mystery comes in when we *look* at the picture ; that is, when we perform the simple act of "seeing," and when we recognise a view as a place that we have been to and should like to visit again. It is the transmission to the mind that is mysterious ; it is there that the reality of the picture lies.

Well now, all this may be regarded as familiar : it may be said, "We knew all that." Very well, I agree, I am only reminding you of what you know ; though indeed it is something of which many people do not think. But I want to use that as a parable of something still more important, an indication of how we shall ultimately have to regard all our perceptions of the Universe in general. The reality lies in the unseen. The seen and the sensed are only dim imperfect representations of it. In many cases the interpretation may still be beyond us : the average man cannot interpret; the scientific man cannot interpret outside his special scope. Interpretation is the function of the Poet and the Seer, the Philosopher and the Saint.

To understand these things we must transcend reason ; not discard it, but transcend it, go beyond it into the realm of imagination and intuition ; feeling our ground insecure, knowing that we are reaching out into the unknown, but

F

yet hoping that in time, and as we develop further, we shall apprehend these things rationally too ; so that reason may once more be dominant. We must be limited neither by our sensory faculties nor by our reasoning faculties. They carry us a certain distance ; they can be supplemented, expanded, stretched ; but they have their limits. The faculty of imagination soars above them. There is a region of mystery even in the most ordinary commonplace action of daily life ; still more is there a region of mystery in the Beyond. But the Universe is not limited to our conception of it, still less to our reasoning about it. We can only patiently grope with such faculties as we possess, and with such aid as has been vouchsafed to us, and thus gain humble access to the region of Faith.

Indirect Realisation.

It is instructive also to realise that many things which undoubtedly exist make a small and intermittent appeal to the senses, or perhaps no appeal at all. I suppose that animals, even birds, are quite unconscious of the existence of the air ; and however much consciousness we might fancifully attribute to the deep-sea fish, we may be sure that they never think of water. That is something like our own condition with regard to the Ether of space, which makes no appeal to our senses ; for even though we utilise its tremors they do not impress our senses as such :

all we apprehend are the objects which by elaborate reasoning and discovery we know must have emitted them.

Even electricity and magnetism make no direct appeal, and accordingly had to be discovered in the course of physical research. Wireless waves are only apprehended by special instruments constructed for the purpose. The air, or rather space, may be full of the result of distant speech or music ; but we go about our work and know nothing of it unless we have put up a special wire to tap it, and have provided ourselves with a telephone and other accessories. In other words, we may be living in a world full of purely physical phenomena, and yet be quite unaware of them. It is only by inference that we know about the interior of the earth, and how much heavier it is than water ; most of it must be as heavy as iron, but no one has penetrated more than a very little way. It is obviously only by inference that we know about the interior of a star, where astronomers suspect processes of the greatest interest, which they have worked out in some detail—processes which never do and never can occur on the earth at all. We have never seen an atom, still less an electron ; yet we study the laws of their behaviour, in remarkable and intricate detail.

A very little additional obscurity of the atmosphere would have prevented our knowledge of other worlds, and of the richly endowed infinitude of space. Our knowledge might easily have been limited to the surface of this planet ; fortunately we have escaped that limitation, though only effectively so in comparatively recent centuries. We realise now what a small object the earth is, and can form some conception of the majesty of creation.

But can we for a moment suppose that we have transcended all such limitations, and that we have become aware of everything that exists ?   Looking back at the physical knowledge of the Ancients, or, if we prefer it, of primitive man, we can see how miserably inadequate their idea of existence must have been.   Surely we can think it probable that our own view of existence is still miserably inadequate !   We are probably unaware of a tithe of what is going on around us. And if a Seer asserts that we are surrounded and continually influenced by a spiritual world, of which our senses tell us nothing—we may doubt the validity of his information, we may decline to accept anything of which we cannot be sure, but it would be irrational and contrary to all analogy to utter a presumptuous denial.

Glimpses of Reality.

As a matter of fact, the evidence is beginning to point the other way ; indeed, to people with a religious instinct, faith has always so pointed. Glimpses have from time to time been caught of some Divine Reality far beyond our sense perception.   Now it seems to me that a claim to

intuition and inspiration is not to be despised.
There surely may be realities of which we can
only catch occasional glimpses, in certain moods
and under exceptional conditions. Many of us
have stood on a Swiss mountain and seen below
nothing but mist ; and yet from time to time, in
patches, now and again, the mist has cleared
away and revealed a landscape of fascinating
beauty far below. Then the mist has closed in
again, the vision has disappeared, and we were
left in what Wordsworth calls "the light of
common day." The reality was there all the time,
but was hidden from our eyes. May we not
take this as a parable, and realise how essentially
limited is our outlook on the Universe ; how much
there may be of which we have not as yet caught
the faintest glimpse ; how much also of which
we do in moments of insight gain some sort of
apprehension ?

But apart from all questions of intuition and
inspiration, the question has arisen in many minds
whether we do not get some indication of the
existence of a spiritual world, through ordinary
prosaic pedestrian methods of scientific inquiry.
Here I cannot speak for all men of science, but
only for some. There is such a thing as psychical
research as well as physical research : and
though the pursuit of psychical inquiry is less
orthodox at present, we must remember that at
one time the pursuit of many sciences which
are now highly respectable was under a ban.
From Roger Bacon downwards, the pursuit of

Natural Knowledge was hampered by authoritative and clerical opposition, its disciples regarded with suspicion, and sometimes ill-treated for their temerity.   Opposition does not stop the investigation, but delays it, and the results have to be stated apologetically.   This is not the place to enter into details ; they have to be studied in Proceedings of Societies now established in many countries, and in the books that serious inquirers from time to time produce.

In matters such as these, since the facts are deep-rooted in the nature of humanity, sometimes the " simple " gain more insight than the " learned."   Too much learning in one direction may clog the faculties in another : things which have not been studied may seem impossible, to those who are aware of their own great knowledge in some direction.   The attitude of open-mindedness all round is not common ; but the facts are there, and are forcing themselves upon general attention.   In time, no doubt, they will be taken under the wing of Science and pursued with the same thoroughness as other subjects.   Meanwhile, I can but state my own conclusions, arrived at through nearly half a century of study.

First, then, I am convinced that the body, which appeals to everyone's senses, is not the whole of us ; that our real meaning and true being are not superficially apparent, but lie in the unseen.   Like any work of art, the physical manifestation is but an indication of something far more deep-seated.

The individual who thinks and hopes and suffers and loves is only in part displayed by the organism through which he manifests. He himself is not the pulpy, highly organised material residing in the skull—which, after all, is only an assemblage of molecules cunningly put together for a time, like a musical instrument or a picture, and which in due course must return to the separate atoms of which it is composed. The body is a temporary assemblage, like a crowd, or a swarm of bees ; the elements of which can disperse, so that the crowd ceases to exist ; but the ultimate reality which underlies it all, which put it together and composed it, does not cease. It may, indeed, go beyond our ken, but it does not go out of existence. Memory does not reside in the brain ; nor does the matter of the brain think and plan and feel. Those attributes do not belong to matter ; they only utilise matter for purposes of demonstration or manifestation. If the machine is injured manifestation ceases. The soul of music does not lie in the vibrations of the air, nor in the instrument which produced the vibrations : it lies in the mind of the composer and in the mind of those who at any future time can perceive and interpret his intention as recorded in matter. A human being is far more than his body ; and there is evidence to show that he has a persistent existence. To those who have studied the evidence, the proof is overpowering that memory and affection, character and personality, survive bodily death. Individuality is developed here in association with matter ; but once developed it does not cease. It loses its own instrument in time, but, strange to say, and fortunately, it does not wholly lose the power of wielding other instruments—though with difficulty and only under proper conditions—and thus it is still able to make itself manifest to those who give it opportunity.

This may be set down as a dogmatic statement : and so it is ; but it is based upon facts. And through those facts it is that some of us have gradually grown to the conviction, here inadequately expressed, that man has a destiny far beyond this mortal life ; that he and his like are beings of high value in the Universe ; that they will be able to strive and serve and influence, long after they have shuffled off the mortal coil ; that, indeed, we are beings for whom any amount of trouble has been, and is being, taken by Higher Powers ; and that it is to this end that all the labour of planetary preparation has gone on. We little know what man may become in the dim and distant ages of the future ; we only know his lowly origin, the root, as it were, from which he has ascended, and the millions of years which have been expended in preparing his home. Depend upon it, these things have an ultimate purpose : we cannot grasp that Purpose in its fullness ; but we can have faith that if and when we do grasp it, we shall realise that all the pain and struggle and effort were worth while : we shall see of the travail of our soul, and shall be satisfied.

Revelation and Vision.

Meanwhile a mere conviction of survival is now no act of faith ; it is the outcome of knowledge : it is established by scientific inquiry. Those who were killed in the war, for instance, full of interest in life and full of affection for those left behind, have given abundant evidence of their continued existence and their unchanged devotion. Being young and unfinished, incomplete in their earth-life, they have had plenty of energy and determination to try to repress undue mourning on their behalf, and to assure their loved ones that they are essentially unchanged.

This effort has always been sporadically made, and has been realised by the few to whom a sense of communion was possible ; but of late years the effort has been specially conspicuous : and if all the incidents could be collected, if even all the incidents known to me were put together, the only refuge from conviction would be a case-hardened attitude of determined opposition and irrational prejudice. The evidence goes on accumulating : we have not to appeal to something in the past which has now ceased : here and now, and continually, the evidence is forthcoming. It is possible to shut our minds to it, to refuse to listen to it, or attend ; but the rejection of fact is not a truly scientific attitude.

The existence of a spiritual world in or around us, and interacting with us, is thus demonstrated. It lies outside our normal ken ; but then, such a heap of realities do that. Essentially, reality can only be inferred, inferred on a basis of fact and good evidence. There is no need to call in the higher attributes of imagination and faith ; the survival of man is becoming an item of scientific knowledge.

If we once overstep the boundary of death, and find that spiritual and individual existence persists on the other side of that transitional experience, then plainly we need dread no other *quasi* terminus. That is why I say the existence of a spiritual world is demonstrated.

Just as we find, on this earth, beings gradually descending in the scale far lower in grade than man—descending without limit to the lowly organised single cell—so may we imagine, in the region above the line of demarcation, grade upon grade of existence above the scale of humanity. We have no reason to doubt the assertion that there are such beings. Even surviving individuals may become, by the continued process of evolution, something far more lofty than anything we know. Grant anything higher than man, and our speculation inevitably reaches on and on, till we come to infinity. Here, in the psychic and spiritual world, is another way in which the Universe is infinite.

Now let us ask what effect on human progress a knowledge of the facts here foreshadowed will have when they are substantiated, when they will be taught to children as a matter of course. There will come a time when every human being will know, perhaps not by his own inquiry, but by universal tradition (just as he knows that the earth is spherical and moving annually round the sun), that the life we see here is only a fraction of the totality of things ; that real life is permanent ; that existence is never-ending ; that evolution and progress continue uninterrupted by the adventure of death ; that only the conditions, and not our personalities, are changed.

In my view, even the conditions are not changed so suddenly and completely as we might without information imagine. Continually we are surrounded by a host of helpers and well-wishers, who are struggling for good, who lament backslidings and weaknesses, who from time to time may be frustrated but who nevertheless persist in their endeavour. They themselves are guided by higher and higher beings, who see and know far more than we do. The whole of existence is suffused with Divine Purpose and Aim, with which in the long run we must get into accord if we are to achieve happiness ; we and they are servants of One in whose service alone is perfect freedom.

All this will in time be known to all. Surely such knowledge must have an influence. In one form or another it has been the message of Religion, the fundamental part of the message, freed from superstitious accretions and sometimes unpalatable details : and surely religion has been a power for good in the world ! So it may turn out that Science, though only trying to ascertain and make sure of fact, yet by the mere power of truth will have an influence for good likewise. Facts ascertained by science are not irreligious at all ; some of them are expressing, in an earth-born and prosaic way, what from time to time has been the inspiration of the Saints and Prophets of all time. These have never felt alone in the Universe : these, whether on mountain-top and Pisgah heights, on sea or on land, or it may be in caves or hiding-places on the earth—aye, and in the very streets of a metropolis thronged with business and crowded with strangers,—have yet felt that they had access to a reality, to a multitude of helpers, a cloud of witnesses, the perception of which was denied to their ordinary senses and to the folk around them. The widespread blindness has distressed them, and they have prayed : " Open their eyes that they may see ! "

Yet if our eyes were often opened we could hardly carry on our daily work ; the brightness would be too great. Any approach to the Beatific Vision may perhaps occur only once in a lifetime. The mists for a moment may clear away ; a gorgeous landscape may be revealed for an instant ; and then the clouds roll back again, and we see no more. But the memory of that vision may be an inspiration for the rest of our lives.

So I think it will be with the majority of men when the discoveries of science are accepted and the meaning perceived. They will no longer be perturbed by wondering what it is all for : they will no longer be overwhelmed by the sense of pain and misery : they will know that these are stages in the process towards some great end ; and some of them will realise that they are privileged to be agents in the process for bringing that end about.

Every leader of mankind, every man who has deeply influenced his generation and has accomplished great acts, whatever the admixture of good and bad in his composition, must have had before him, perhaps frequently, perhaps only at times, some sense of the Divine Purpose and Mission entrusted to him. Men of thought, like Socrates and the Philosophers ; men of imagination, like the great Poets ; men of action, like Cromwell and Abraham Lincoln, not to mention some in our own day,—aye, and men of business too, who without much show and demonstration have devoted themselves to the betterment of humanity in such directions as come within their scope,— all these, in one form or another, must have been inspired by a feeling of responsibility, by a realisation that there must be some worthy outcome, in the fulfilment of which they were privileged to share, for which all their efforts were asked and needed. Such a faith adds greatly to human power : without it the individual is helpless and alone ; with it he can do his duty. And those who have that faith in large measure are the helpers of the race.

The way to the Vision and to the Reward, " Well done, good and faithful servant ! " lies through the full exercise of our talents and through hard work.   So

> . . . lay thine uphill shoulder to the wheel,
> And climb the Mount of Blessing, whence, if thou
> Look higher, then—perchance—thou mayest—beyond
> A hundred ever-rising mountain lines,
> And past the range of Night and Shadow—see
> The high-heaven dawn of more than mortal day
> Strike on the Mount of Vision !
>
> So   farewell.

FAITH AND THE QUEST FOR TRUTH

RELATIONSHIP BETWEEN THEOLOGY AND SCIENCE.

IN trying to get some mental grasp or partial comprehension of the Universe of which we are a part, it is most important and helpful to realise that we are a part of it, and that all the parts cohere together into a unity. The human mind is evidently something not alien to the Universe : it is in harmony with it ; and its instincts and its judgments, so far as they go, may be trusted. We do not always know on what grounds we judge, or on what basis our opinions are founded ; but instinctively we discriminate between the beautiful and the ugly, between the noble and the unworthy, between the good and the bad. We may make mistakes ; individuals may obviously be mistaken ; but the general trend of the human mind is in one direction, and often we have no better guide than our instinctive feeling.

We may extend the same general notion, away from Ethics and Æsthetics, to the pursuit of Truth. Imagination and inference must far transcend observation ; in my last lecture we took pains to realise how little direct observation gives us, and that existence is not in any degree limited to what we perceive. We are constantly perceiving things not yet understood : we often do not realise how little we know about them. An enlarged perceptivity can give us a multitude of truths which we cannot yet grasp and formulate, but about which we have faith that they will be ultimately intelligible, and come within the scope of Reason. It is, indeed, our experience of the *consistency* of Truth which has kindled the light of reason in our mind and given us confidence in its all-prevailing efficacy. The more we understand of things, the more satisfying they are : we have never yet been let down, or confounded, by anything that was really true. Truth is always a step up. Some truths are many flights up ; and we may proceed in the hope that we shall arrive at them in due time. "O Lord, in Thee have I trusted : let me never be confounded."

Reason carries us a long way, and it is a faculty to be thankful for ; but we should ill-treat it if we regarded it as a limiting faculty necessarily excluding things beyond it, even if they seem at present beyond its possible scope. The human mind is progressive ; its scope gradually opens out. Things that come within its purview are often first dimly grasped by faith, then realised in imagination, and finally understood by reason ; and every advance, by increasing the boundary of knowledge, tends to enlarge our scope further. The use of reason was not for purposes of negation. A person who calls himself a Rationalist seems to forget

that, or not to admit it ; but I repeat—reason
is not for purposes of negation, but for purposes
of enlarged and more complete comprehension.
Perceptivity must go far beyond reason, and
reason should not deny that which has not yet
come within its scope.   To deny anything effectively demands superhuman knowledge : it is
much more presumptuous to deny than to assert.
Assertion may be based on experience : we cannot
have experience of negation.   Wholesale denials
are not rational ; they are hypothetical and
presumptive.   They are often the result of pre-
suppositions and prejudice.   Major premises,
whether positive or negative, are the result of
induction from certain instances.   They are
hypothetical generalisations ; they may not be
true ; a single clear opposing instance would
upset them.

For instance, the generalisation or major
premise, " All swans are white," can be upset
by a single instance of a black swan.   But it
would have to be established that the bird
really was a swan, and that it had not been
artificially blackened.   So there would be room
for argument and scepticism.   The generalisation, " The dead know not anything ; in that
day their thoughts perish," or, in modern
language, " The dead are extinct, non-existent ;
their activity has wholly ceased," can be upset
by a single authoritative message establishing
the identity of a deceased communicator. ·The
establishment of identity under those conditions

is admittedly not easy ; and many other hypo-
theses would have to be tested and discarded as
inadequate before the fact could be admitted.
Moreover, it would be very rash to base anything on a single instance : evidence must be
cumulative, and only after long and patient
experience should a judgment on any really
vital proposition be formed.

The existence of discarnate spirits can only be
proven by their producing some physical effect
or by our entering into communication with
them. That is how the existence of denizens
of Mars could be established, if there are any
intelligent beings there. We have not got into
touch with Mars, and I doubt if we are likely
to, but in my judgment we have got into touch
with the discarnate. It is not a light and easy
thing to say, but there is nothing irrational
about it. Consider the reaction of a race of
savages faced with strange phenomena which they
could not understand. Most of them would take
refuge in denying the phenomena ; some might
accept the facts without theory ; but others might
suggest that they were due to the activity of an
unknown race of white men. This last could be
stigmatised as wild superstition, but it might be
true for all that. What we have to do is to find
not the pleasantest or most plausible theory, but
the true one. If it be true, its strangeness will
wear off in time ; it will be recognised as completely harmonious and fitting into its place in
the general scheme.

Rationality of all Phenomena.

Depend upon it that the Universe is rational ; we should have faith that everything that is true will ultimately be recognised as part of a rational universe, and that there is no ultimate discordance between existence and the human mind : the two are in final harmony, though we at present comprehend so little of the totality of existence that we may be unduly sceptical about phenomena which in due time we shall understand. Some facts occasionally seem absurd at first : facts often do when we do not understand them. Regions of half-knowledge may be full of confusion and absurdity until we are able to rationalise things by completer knowledge. But to receive such knowledge we must have an open mind. A mind which shuts things out of consideration on the ground of apparent absurdity is not a mind to make new discoveries with, nor is it a mind in real harmony with the Universe. Not so many years ago it would have savoured of superstition to suppose that a gauze screen could keep off malaria, or that pulp from a snake's head could cure, or render anyone immune from, snake-bite. Galileo even regarded it as fanciful to suppose that the moon had anything to do with the tides. The Government order at Panama that every discarded tin vessel should be punctured might be made to sound like an insane superstition ; and some of the practices of savages may after all be justified, in

ways which at first, to civilised man unacquainted
with all the conditions, seem preposterous.

All this is conspicuously true of the phenomena
studied in psychical research : many of them
appear absurd and quite incredible ; but, in so
far as they are true, their unreasonable character
can only be due to our ignorance. Orthodox
science may have a difficulty in accepting them
until they become rational ; but in human
history mankind has had to accept many facts
which it has not understood. Nothing will be
ultimately inharmonious with human reason, but
many things seen at first through a mist of
ignorance may loom large and ugly and incredible, until they are better understood.

The domain of Theology, again, lies for the
most part and at present outside the reach of
human reason : it has to be apprehended by
other faculties. Faith is there our guiding light.
Faith is quite different from credulity, but still
it is a faculty often looked at askance by scientific
men, though really they are constantly exercising
the faculty in unsuspected ways. How else
could they draw their multitude of inferences ?
The very Uniformity of Nature, on which science
depends, can only be an article of faith. Faith
rules the daily life of us all. Faith, under the
name of credit, is the sheet-anchor of business
men. Naturally faith is essential to Theology.
The whole conduct of life is governed by faith.

Theology lies outside the scope of these
lectures ; or if it does not, it will doubtless be

treated henceforward by some Theologian. Meanwhile the pursuit of science has had some influence on Theology : and the view which humanity can take of the ultimate Being must be either helpful or hindering to its progress. The Bishop of Birmingham has truly said that, though we are occasionally baffled by the occurrence of the unexpected and unpredictable,

the progressive development of scientific thought emphasises the unity of plan and structure of the Universe. The cosmos hangs together. The bundles of regular sequences in it which we have discovered indicate the existence of true cause, of purpose expressed in a single ground-plan. It seems to me that the existence of some sort of a God, to whose Intelligent Will the whole is due, has become an increasingly reasonable belief.

But it must be admitted, as he admits, that this does not carry us very far. Reason is constantly stopping short at the boundaries of the known, and leaving us to proceed further by other faculties if we can. It is surely reasonable, for the majority of us, to be sure that the intuition and genius and inspiration of the higher few—the saints—has carried them far beyond our pedestrian contemplation ; and we may surely hope that from those visions they have emerged and told us what we, too, shall ultimately find to be true.

In that way the Christian faith holds that God from time to time reveals His Nature to man, and that in spite of, or perhaps because of, His inscrutable majesty and power, He is yet surprisingly accessible, having been revealed as the Loving Father of Christ's Gospel, Who cares not only for the host of heaven and for humanity, but for the smallest creeping thing upon the earth.

What I might venture to call one of the most unfortunate sentences in the Bible—I mean in the evil effect it has probably had—is the question, " Doth God take care for oxen ? " (1 Cor. ix. 9). In itself it is a mere piece of forensic pleading, to sustain the argument that everything was planned for humanity ; but it has been stretched, perhaps unconsciously, so that in Latin countries the Church does not seem to regard it as its corporate business to protect animals. It has been said " they are not Christians," and they are too often ill-used. Does God care for oxen ? Yea, verily He does ; but their welfare and that of the domestic animals has been entrusted to us. The sparrows and wild birds and beasts are less directly in our care : " The lions roaring after their prey do seek their meat from God." Surely an all-embracing Love and good will is a Divine attribute, as well as Goodness, Beauty, and Truth. The association of pleasure with simple acts, the convenience and the amelioration of life resulting from every advance in knowledge, are instances of this. These things come unbidden, and so to speak unsought, not primarily sought. Our aim in science is truth, not practical utility. Utility may come as an extra ; it may be added unto us, but it is not that which we seek first.

" The Universe is the outcome of Supreme Mind : abandon this faith and the only value of Science is its practical utility." How many people there are who only think of science in terms of practical utility and of the daily conveniences which it has made possible ! Yet humanity as a whole is beginning to know better. There would seem to be very little practical utility in the study of the stars : there can surely be none in the theory of relativity ; yet these subjects have excited wide interest, have been taken up by the daily Press, and are lectured upon in all the continents.

Instinctively, and perhaps at first irrationally, we feel that a wider and truer and more comprehensive outlook upon the Universe must be good, must conduce to fullness of life, must be a worthy subject of interest and study, must in the long run influence our outlook, and perhaps have a practical effect upon our conduct. We cannot really tell in what direction any new truth will be applied : we must have faith in truth for its own sake, and realise that truth is so big a thing that it, too, may have various aspects and different modes of manifestation.

Varieties of Truth.

I suppose it may be truly said that whether in Science or in Religion we are all out for truth ; in fact, nothing else is of any use ; but absolute Truth is difficult to come by. The Universe is a very big and comprehensive thing ; humanity is, after all, only a corner of it, and it is not surprising that there are, and always have been, differences of opinion as to what truth is. Full and absolute truth must be beyond our grasp : our capacity is insufficient to hold it. We must be content with such aspects of truth as we can perceive, and if some of us perceive different aspects, there is no need to quarrel on that account. All we can do is to be faithful to such truth as we do perceive, and to seek for more.

It will be generally admitted that there are different kinds of truth. There is a truth of Literature, which means, I suppose, something true to human nature—quite a different thing from true to historic fact. This is the truth which dominates *Hamlet*, *Othello*, and *King Lear*. No one asks, or need ask, if these people really lived and acted in the way represented. It is a platitude to say that the truth of Poetry differs from the truth of scientific fact. If a drama is true to human nature, that is all that is required of it, at least in that direction.

How effete all the controversies about Genesis and Mosaic cosmogony become when this is clearly apprehended ! The great drama of evolution cannot really be conveyed to humanity even now ; still less could it have been apprehended in primitive times. It can only be expressed in terms of drama and poetry. " God spake." " Let there be." " The Spirit of God brooded over the face of the deep." "In the beginning was the Logos "—the Divine scheme or plan. "Without it was not anything made that was made."

These are the realities, not the form in which the incomprehensible drama of Creation is drawn by poet and scribe. "Male and female created He them " is a truth deeply embedded in nature, not the imaginary and pictorial details. The six days, and other trivial accessories, belong to the realm of literature, not to the alien region of prosaic fact.

Then there is the truth of Art, which in its highest form seeks to dive down into the soul of things and represent an ideal which otherwise might be beyond our ken, beyond our sense of perception, beyond the scope of our normal faculties. Of this nature perhaps is the truth of inspiration, of intuition, of something physically unperceived and yet mentally or spiritually grasped ; grasped perhaps only by a genius in moments of insight, and by him made partially manifest to others. Many things are too big for verbal utterance ; not everything can be formulated in words. Fine music, for instance, can awaken a sense of cosmic perception not easy to recover, not easy to convey. Saints have been caught up into the third heaven, and in a period of ecstasy have realised things which it was not lawful or possible for man to utter. The Universe is big enough to contain all these and much more. It cannot be grasped in its entirety by human intelligence ; its fullness is not limited by human capacity. The possibilities in the totality of existence must be overwhelming and appalling. Only here and there, and occasionally, can we hope to express what is true.

Then there is the truth of Science, which at first sight seems limited to actually ascertained fact. But science deals with a multitude of things which are not directly apprehended by the senses. It, too, is full of intuitions and inferences and guesses and theories. It is based upon sensible fact, but it transcends all direct apprehension, and launches out continually into the Unknown.

There is some analogy between the truths of Science and the truths of Religion. Both have been gradually arrived at by a process of evolution. Our early ancestors were incapable of more than a fragmentary conception of either. Their ideas at one time can have been but little above the animal's. The elaboration of such ideas into Theory, or Dogma, has been a gradual process involving many intermediate stages ; and, in spite of the confidence with which those ideas have been promulgated and the tenacity with which they are held, they are not likely to be anything like complete. Far from it. The Universe is infinite, and therefore not to be wholly grasped by Science. The Divine is surely still more infinite, and therefore Theology can but make inadequate attempts to express it.

Both Science and Theology are based on what are considered to be facts ; both reach far out into the unknown, the superhuman, the awe-inspiring. Science is concerned both with the infinitely small and with the infinitely great, with the operations of life and its interaction with matter, with problems of infinite duration and cosmic significance. Surely Religion is, too ! Both at times tend to consolidate or fossilise into dogma, instead of retaining their birthright of fluidity and freedom.

LIKENESSES BETWEEN SCIENCE AND THEOLOGY.

I trace a certain analogy in the history of the development of Science and of Theology. Both are above the heads of the average of mankind, yet Theology in its religious aspect profoundly affects the beliefs and behaviour of every man ; and similarly Science, in its applied aspect, has had deep-seated sociological significance. So the reactions of mankind to Religion and to Science have been not altogether dissimilar.

It may be said that Science has no system of rewards and punishments, as Religion has. But wait. Science recognises that misdoing will be appropriately punished, not by arbitrary decree, but by inexorable law. Its punishments are even more severe, less mitigated by compassion and kindliness, than are the punishments of any truly Christian Religion. As Huxley said, the method of Nature is not a word and a blow and the blow first, but the blow without the word ; you have to find out for yourself why your ears are boxed. To restore the balance, one School of Theologians invented a savage system of punishment which put Nature into the shade ; and while they were about it, there was nothing to prevent their continuing the inhuman treatment to all eternity ! What wonder, then, if the theories of Theology tended to harden into fiercely expressed dogma, which should put the dogmas of Science into the shade, and that Theologians should require from devotees a whole-hearted assent to propositions whose mystery and incredibility only added to their value as a test of submissive credulity. Science has done its best, but never has it been able to approach the virulence of theologians in insisting on detailed doctrine and punishing all those who in the smallest degree departed from it. Science never had the State behind it ; it was unable to hand over recalcitrants to the secular arm. Science is not wholly free from priestcraft, but it is comparatively mild in its methods of enforcing its decrees. Ostracism and ridicule are its negative methods : milder these than the positive methods of the thumbscrew, the rack, and the stake ! While as to punishment in the next world, militant science has hitherto contented itself with a negative, saying virtually : Be not befooled with the notion that you will outlive your bodily manifestation, for that is the only " you " that exists. People on the other

side of the veil cannot possibly communicate
with you ; the idea is absurd, for there are no
people there. That is what they say. Intelligence and Memory cannot persist, for they are
limited to, and are merely functions of, the
material brain. What a monstrous assertion !
What an unfounded hypothesis ! The brain is
needed for the apprehension of these things by
our material senses—true ; if the brain is
damaged the manifestation is impaired—quite
true ; and if the brain is destroyed its function
as an intermediary between mind and matter
ceases.

Certainly, all that is familiar fact, but what
then ! Can we on the strength of these facts
concoct a major premise and say that no mind
exists save in association with a material brain ?
We might be tempted to suspect even that, if
no facts clearly indicated the contrary. In
face of those facts the absolute dependence of
mind on brain is a presumptuous generalisation, unworthy of the cautious attitude generally
characteristic of scientific men. The attitude of
Theologians is different, but they, too, as a group,
have a complete system, and dislike even their
old traditions in a new garb. They resent proofs
which savour of novelty. Occurrences appropriate to many centuries B.C. and to one or
two A.D. seem inappropriate nineteen centuries
later. The facts are objected to on the ground
of theory—or rather on the ground of prejudice
and presupposition, though the prejudice and

presupposition are by no means the same as those of the men of science. Theologians and Scientists may differ on most things, but they agree in ostracising and condemning unusual psychical phenomena.

So there are differences. But differences are obvious ; it is similarities which may escape attention. Truth is the basis of Science as well as the basis of Religion, and the history of some stages in arriving at truth contains some instructive likenesses. I will not emphasise the likenesses in these stages ; each can do that for himself. I will only run over the history of Science from this point of view, as a sort of parable or parallel of what might by another better informed person be run over as the history of Religion.

In Scientific history, then, I note, and presently expound, three stages :

(1) An era of what might be called literalism, or fundamentalism, or the scientific equivalent of bibliolatry ;

(2) An era of orthodoxy, ecclesiasticism, and established doctrine ;

(3) An era of emancipation, freedom, development, an attempt to restore fluidity, to test and try everything, and to hold fast that which is good.

Are not these stages manifest also in Theological history ? In the latest, more recent, more hopeful, Modernist era nothing is too sacred for reverent reconsideration—any other kind of reconsideration applied to that which has been the faith of thousands would be unmannerly, ill-bred, and unworthy—but, with due reverence for the old, no ancient formula is to be treated as infallible, everything is held subject to the guiding spirit of free inquiry.   Oaths and binding declarations ought not to be thoughtlessly administered, so as thereafter to enshackle the soul and constrain it to give up the search for truth and degenerate into an unrepining state of fixed and unprogressive orthodoxy. Exaggerated protestations, such wholesale exaggerations as "With all my worldly goods I thee endow," "All this I steadfastly believe," and other thoughtless declarations inflicted on the faithful in moments of emotion, should be deprecated and gradually eliminated.   All these shackles are to be thrown off.   The free air of heaven, the Spirit that bloweth where it listeth, is to be welcomed with upturned face and clear eyes and abundant health.   While the Spirit which constantly denies—Goethe's Mephistopheles "*der stets verneint*"—is to be relegated to the murky prison from which the shackled guards have let him escape.

FREEDOM AND FLUIDITY.

Freedom and fluidity are the life-blood of Truth.   Consolidation is stagnation, and the limb goes to sleep.   Fixity is not faith.   Dogma is said to crystallise out of the fluid from time to time and to form a deposit of faith ; but solid deposits, like calcareous concretions, are injurious to a living organism. The deposit of faith can be nothing but accumulated tradition, very worthy of respect, not lightly to be discarded, but it should be plastic, not crystalline. It is not really eternal, immutable, infallible ; and if so treated it will become an obstruction and check the flow of the life-blood. Articles of Belief may be imposed by orthodoxy ; but faith can never be imposed. The Universe is too vast, too majestic, to be comprehended and formulated by the creatures on one of its planets. Both religious and scientific organisations are liable to forget this, and virtually, though not always consciously, insist on certain articles of faith in a spirit of rigid orthodoxy, which is inevitably opposed to the spirit of free inquiry. It seems to me that the attempt to curb the spirit of free inquiry is responsible for the dearth of candidates for the Ministry. Earnest youth recognises that a minister is officially bound to teach certain doctrines, and of that the public are well aware; so their utterances are liable to be discounted, even when quite genuine. Moreover, if ministers depart from those doctrines in the least degree, and hint at something they have realised or assimilated for themselves, an outcry is raised, and they are accused of heresy or of infidelity. The Press publishes these utterances, and correspondents accuse them of professing one set of opinions and holding another. They are not really free, or at any rate their freedom is limited, sometimes by ecclesiastical authority, sometimes by their congregations and the public.

The possibility of all this may deter promising students from entering the Ministry. A conscientious youth may hesitate to load himself with chains, however elastic in practice those chains may turn out to be. A speech-restricted priesthood neither feels free itself nor is it regarded as free by the laity. Can they be other than " cabined, cribbed, confined "—perhaps not in their search for truth, but in their promulgation of it ? Too well are the creeds of the various denominations known and verbally insisted on, by the stupid, the vulgar, the illiterate, and the well-meaning. Some take refuge in ceremony and " services," but their chance of holding the attention of the men and women who are doing the work of the world seems small. Some feel so bound by what they have professed that they give up further study. Some are hampered in their further study by lack of leisure, and even by poverty. There are plenty of difficulties. Yet many overcome them, and stand up for truth and freedom. In spite of difficulties, the interaction of divine and human is preached with real conviction ; men rise superior to their artificially imposed limitations.

Outside any Church, Brotherhoods of working

men are active and mutually helpful in a truly Christian and most hopeful spirit. Inside the Churches there are hopeful movements. Religion is not a matter of intelligence alone ; yet if it be not intelligent it will not survive. The intellectual is not all, the pastoral is not all, the sacramental is not all. But there is room for all. There is truth and possibility of development in all these things. The spiritual and the material worlds do interact, and when their interaction is more fully recognised I see possibilities of further progress and hope of the coming of the Kingdom.

SKETCH OF SCIENTIFIC DEVELOPMENT.

I shall now take up my parable and trace the history of scientific development under the above three heads—the traditional, the orthodox, the free ; and incidentally shall continue what I was saying about the truth that even Science is not limited to actually ascertained fact. Science deals with a multitude of things not apprehended by the senses : it, like Theology, launches out continually into the unknown.

Looking back over the history of Science, we can trace several stages in scientific development. For many centuries the world was dominated by the Aristotelian philosophy, a tradition handed down from generation to generation, based largely upon ancient writings, which in the course of time acquired a sort of infallible authority. Assertions were made, and not questioned : there was a fundamental body of doctrine to which appeal could be made, and any departure from it was heresy.

In due course, however, heretics arose. That rather mysterious figure, Roger Bacon, was a pioneer far in advance of his time, and was persecuted accordingly. Much later, and in a more congenial atmosphere, Francis Bacon eloquently advocated the right of private judgment, a method of direct appeal to experiment, a claim that each individual could make observations and determine truth for himself ; with the implied further claim that if experience ran counter to fundamental tradition, then it was tradition that must give way. The chief heretic to put this into practice, of his own motion and independently, was Galileo ; and, like all innovators, he suffered the consequences of his temerity. But the new method of arriving at truth conquered in the end, and in due time attained a kind of orthodoxy and supreme authority, under the splendid reign of Sir Isaac Newton. He and his co-workers and disciples grasped scientific truth to an extraordinary degree. It was felt that at last the Universe was comprehended. Not indeed was it so felt by Newton himself, but by his followers, who considered that the main foundations of science were now laid down, and all that remained was to build it up in detail, elaborating the structure with surpassing ingenuity and mathematical skill. Under that scheme we have

lived ; the vast progress of the nineteenth
century is built upon it ; and fifty, nay thirty,
years ago it seemed as if nothing could disturb
its magnificent serenity.   It could be added to,
supplemented, developed, continually increased,
but always along the same fundamental lines.
It was felt that the main outlines of the Universe
were understood, and that there could be nothing
so new and different as to be revolutionary.
The atoms of matter were there, had been there
from time immemorial.   Matter and force reigned
supreme.   Everything was subject to the laws
of Dynamics ; and, as Laplace said, given the
motions of all the particles, the past could be
reconstructed and the future foreseen.   The
Universe consisted of atoms, perfect in number
and weight, unchangeable and permanent.   The
laws of their motions were known, and the theory
was certain.

But in our own time a new heresy has arisen,
a kind of Modernism which calls all this orthodoxy
in question, which pulls up the roots of long-
planted ideas, examines them, and finds them
deficient ; finds some of them, indeed, imaginary,
unnecessary—human interpretations superposed
upon the reality of things.   The effort has been
made, therefore, to get deeper down to reality,
to perceive that much which had been thought
permanent was changeable, that the vast continuity of the forces of nature was interrupted
by a kind of discontinuity, and that all our most
cherished ideas must be re-examined, recast and

perhaps reformulated. Even matter itself was perhaps not permanent, but could pass away into energy ; force might be non-existent ; the consequences of the laws of dynamics were only approximations ; everything must be put once more into the melting-pot.

The younger heretics of our generation, in order to emphasise the truth that they are now enthusiastically advocating, sometimes seek to pull down Newton from his pre-eminence, to look behind and re-examine once more the foundations of Science, and to develop a new system of law and order, a new method of formulation ; not in their view merely supplementing, but in some respects discarding and re-creating the old. The creed of Science is being reconstituted. Old traditions are uprooted ; and a living spirit of inquiry, which shrinks from nothing, is evolving theories which must indeed be tested by experiment, but are developed much more by the powers of the mind and the inspirations of genius than by any actual observation or sense perception.

For the things now dealt with are not sensible realities, or at least they are so suffused with new ideas that their experimental basis is mainly of a negative character. Certain results which were expected on the old view have not materialised ; and accordingly a so-called theory of Relativity has entered on a triumphant career. It is a quest for Absolutes. The theory proclaims a new and surprisingly small list of

absolutes, in the midst of a wilderness of humanly
interpreted and mainly relative phenomena, dis-
lodged from their late pre-eminence. The inde-
structibility of matter, which was the fundamental
basis of chemistry, is no longer believed in.
Even the conservation of energy is sometimes
doubted. Force, the basis of dynamics, is held
to be non-existent ; and the physical universe is
explained, or is attempted to be explained, in
terms of a transcendental geometry not given
by experience at all. The fundamental abstractions of space and time are laid violent hands
upon, and are treated as human illusions. Even
the locomotion of matter, the most familiar
thing in our ordinary life, is found to be something other than appears. And what the end
may be, who is to say ?

In this revolutionary age, some may be going
too far ; others in all probability are not going
far enough. Aristotelian Fundamentalism has
long been discarded ; the quasi-ecclesiastical
orthodoxy of Newton is being overhauled : new
ideas are everywhere dominant. The permanence of the everlasting stars is questioned ; the
birth and death of worlds, their ages, and the
processes by which they have come into being,
are being calculated and critically examined.
The process of evolution is being studied
throughout thousands of millions of years.
More is known about the interior of a star than
about the interior of the earth. The constitution
of the atom of matter is unravelled. The nature

of energy and that of the ether—if, indeed, there be an ether—are becoming supreme questions.

And what is the result, so far as we have gone at present ?

First, an enlargement of the Universe beyond all previous conception.

Next, a detection of law and order running through the whole ; the same laws ruling in the farthest star as in this little solar system ; a constant flux and activity, constant and inexorable ; change and development, with a possible recurrence and perpetuity of an unexpected kind in the material universe.

There is no impiety in the proceedings. Humanity is stretching its powers to the uttermost, and is, I believe, unconsciously preparing the way for a perception of spiritual truth which has not yet fully dawned on the scientific horizon, but of which I catch the glimmerings that precede dawn.

Can it be wondered at if all this scientific upheaval must have a bearing on the problems and the kind of truth appropriate to Theology ? The old interpretations, the old formulæ, the old traditions, will have in their turn to be re-examined, reinterpreted, and perhaps reformulated.

REACTION OF TRUTH ON FAITH.

Aye, and we must go further—further than is as yet contemplated. The Universe, as I believe, contains the substance of new and unexpected discoveries. The physical universe and the spiritual universe are interrelated : we are only beginning to recognise the possibilities of their interaction. We know that they do interact, by direct human experience : the familiar action of the mind on the body is sufficient to show that. But the constant activity and the full attributes of the Spirit have yet to be revealed. Some of us are exploring in dark corners and working through a jungle into clearer light. We are always instinctively careful about what we assert : my own view is that in these revolutionary times we must be scrupulously careful about what we deny. I believe that the old will be supplemented and improved, rather than destroyed and discarded. The True will remain. Science comes not to destroy, but to fulfil. The Universe is as it was, and as it is, in spite of our temporary and plausible discoveries. The older generations had their avenues to truth, just as we have : their truth was not complete, neither is ours. Let no generation ever think that they are coming to an end of discovery, that their facts are final and unassailable, that their theories are complete. Infallibility is not for man.

In particular, and as an example of what I mean, if I trespass off my ground and on to the ground of the Theologians, I want to say that, as far as I can judge, the progress of Science is tending towards a strengthening of

Theology in all its really vital aspects ; and that certain narrations which have been doubted— I shall be understood by many here if I cite as examples the direct voice at the Baptism, the Presences at the Transfiguration, the vision on the road to Damascus—were true happenings. True, that is, not merely because of historical evidence, about which many are better judges than I, but because things like them *can* happen. And I look to the time when the constant interaction of spirit and matter will be more fully recognised ; when the term " spirit " will be extended to human spirit, and the Incarnation can be rationally recognised as both a Divine and a human fact. The Divine and the human are truly interrelated ; they do interact ; the spiritual world is a reality. The reality of a spiritual world is the sheet-anchor of religion : it is being demonstrated by science. The spirit of man is but a fraction, a minute fraction, of the Universe ; nevertheless it *is* a fraction of the totality of things. And the totality of things, as we apprehend it, is but a fraction, and yet a real fraction, of Divine Reality.

Here is a poem whose source I know not ; I found it somewhere recently :

Lord, I believe !
Man is no little thing,
That, like a bird in Spring,
Comes fluttering to the Light of Life,
And out into the darkness of long death.
The Breath of God is in him,

And his agelong strife
With evil has a meaning and an end.
Though twilight dim his vision be,
Yet can he see Thy Truth.
And in the cool of evening, Thou, his friend,
Dost walk with him, and talk
(Did not the Word take flesh ?)—
Talk of the great destiny
That awaits him and his race
In days that are to be.
By grace he can achieve great things,
And, on the wings of strong desire,
Mount upward ever, higher and higher,
Until above the clouds of earth he stands,
And stares God in the Face.

As says Gilbert Murray in a recent *Hibbert Journal* :

We know that we can find, not all truth, but some truth. We know we can create beauty. We know we can live, or try to live, in the pursuit of goodness, making our will the fellow-worker of God.

Faith is the substance of things hoped for, the evidence of things not seen. Aye, truly, faith is the motive-power of humanity. By faith we make mental inferences. By faith we regulate our lives. Had we no faith in the future we should become supine. Had we no faith in goodness and love we might well despair. By faith we strive, we hope, and up to our measure we achieve. Through faith, though at times we suffer, we cling to a larger hope ; we reach a hand through time to catch the far-off interest of tears. Already we can begin to make preparation for the ultimate unknown destiny of man.

LIFE AND ITS MYSTERIES

SOME INSOLUBLE PROBLEMS OF EXISTENCE

By " insoluble " problems I mean those to which
we have as yet no complete or satisfactory clue.
They will be solved in good time. Some show
signs of incipient solution already.

The first problem that I shall tackle is the
Problem of Evolution. Why should there be a
gradual process at all ? Why should species
arrive which turn out unsuitable and fail ? Why
should there be a struggle for existence and a
survival of the fittest ? Why should anything
come into being that is imperfect, or less fit
than it might be or than others are ? Why
are there malformations, monstrosities, and utter
failures ? These strange occurrences give to
the process an air of experimentation, like a
tentative operation, as if there might be gradual
improvement on first attempts. This seems a
very human method of creation ; it is just an
instance of how our minds are in harmony with
facts. Nevertheless, to primitive races, such a
method might naturally seem unworthy of a
Divine Creator. That is why, I expect, or
partly why, Evolution is still objected to by so
many : it is so different from the sudden production of full-blown things in an instant, at a word,

the " Let there be, and there was." That seems dignified and stately. And so, poetically, it is. Poetically it is true ; only the element of time and all the intermediate steps are left out. The elimination of time, the skipping of intermediate steps, is the method of magic, the method appropriate to the ideas of early man. Remember the stories of magic in the *Arabian Nights*. You rubbed a ring or a lamp, summoned a supernormal power, and straightway a palace was erected in a flash. Sometimes the germ of an idea comes like that, in a moment of inspiration. " The flash of the will that can," as Browning called it. But in the natural and physical world of actual life we do not find things happening in that way.

It is perhaps conceivable (though I doubt it) that this *might* have been the method of creation, but by reverent examination we find that it was not the method employed. The actual method was a slow and gradual one, involving great tracts of time. No doubt this process is in truth entirely reasonable ; but whether we know the reason or not we must accept the fact that so it is. Always be loyal to facts !

But shall we therefore, on the strength of admitting evolution, deny that there is any design or plan behind it ? Shall we speak of it as a self-acting, a sort of random, process ? That would be folly. The products of human art and industry, like everything else, even steam-engines and dynamos, looms and printing-presses, drugs and explosives, have arisen in the course of evolution : they were not brought into being suddenly ; they are the survivors of a number of attempts. So a misguided person, yet speaking truly up to a point, could say, looking at the machinery in a factory :

> Of course all these things work properly, and look as if they had been designed for that end ; but the fact is that all those that did not work properly were scrapped. When something is hit upon that works well, it survives ; if it is unsuited to its surroundings, it soon perishes. That is how mankind acquired the wheel, the boat, the steam-engine, the telephone, and the aeroplane. To say that the things we utilise now are the result of some human design at the beginning is false : no one could have conceived them as they are now. They have arisen gradually in the course of evolution by the survival of the fittest. The structure of an aeroplane, or even of a bicycle, is the result of trial and error ; the present elaboration is due to a series of blind attempts at putting things together : there was not at any time any mind in the process. The stages can be followed in a museum, like fossils in the rocks ; the gradual development of every automatic machine, and of every self-acting organism, is manifest.

That is what might be said about the devices constructed and used by us, on the hypothesis that we were automata, mindless automata, through whose agency nevertheless things came into being that looked as if they had been designed.   Biologists will object to the parable ; but it seems to me cogent enough.   It is, of course, *untrue* that mind was not active at every stage in the process ; but in the parable that fact has been overlooked or denied.   What is true is that the ultimate development and elaboration of (say) the telephone or the aeroplane was not conceived at the beginning.   The *Majestic* was not thought of when the first boat was hewn out of a tree-trunk.   Perfection is only reached by a number of intermediate and imperfect steps.

The theologian might put in a forcible plea that the impossibility of foreseeing on the part of a human being need not apply to a Being who can look before and after ; who has available all the experience of innumerable other worlds, if indeed experience is necessary at all ; and who might reasonably be supposed to foresee even the ultimate results of a long line of progress.   Indeed, not only the ultimate results might be foreseen, but the intermediate stages of incompleteness and frustration and struggle and destruction might be foreseen also, because for some unknown reason they appear to be inevitable concomitants of an agelong process.

It may well be asked, by both Scientists and Theologians, why these intermediate stages were necessary, why duration is so important, why there should be an appearance, perhaps a reality, of struggle and effort, of partial failure and gradual success, on the part of what Bernard Shaw calls the Life-force ; and we may not be able to give a satisfactory answer. But surely it by no means follows that a satisfactory answer could not be given to one who had sufficient knowledge to receive it. The faculty of reason, as so far developed in humanity, is often brought to a standstill against insoluble problems, since our reasoning power is still in a nascent stage. We have only experience to show us that so things are ; and by faith alone do we surmise that so they must be.

Nevertheless, in speaking of and recognising the appearance of struggle and effort, we seem to be speaking of and recognising something real, something genuinely and fundamentally incorporated in the nature of things. Results of high value can perhaps only be achieved through blood and tears. I believe the effort to be not artificial or gratuitous, but necessary ; the work of creation may be really difficult ; it is unlikely that things can be produced " easily by a nod." Certainly it would be a mistaken admiration that should attribute ease of production to a human genius. A life of value is always a strenuous one, full of effort, and often of pain. So also it is unfair to suppose that the production of a human race, endowed with free will—a race which with the power to go wrong shall ulti-

mately, of its own volition, determine to go right—
is an easy task. The effort *humanam condere
gentem* is real and vivid, the call upon us for help
and co-operation is no artificial piece of play-
acting, but is deep-rooted in the nature of existence. Perhaps we cannot reason out why it
should be so ; but surely we can learn from
experience that so it is, and may have faith that
some day we shall understand it better. Already
there may be some who think they partially understand. " My Father worketh hitherto, and I
work," is a saying attributed to One with greater
intuitive knowledge than any other of the sons of
men. We may take it that it contains a real
truth.

Many have been the attempts to solve the
theological problem, and equally numerous have
been the failures. One of these attempts has
been to postulate a principle of evil, in controversy
with and opposition to the principle of good—
an evil agency which is able to undo good work,
to exert opposition force, to offer active resistance.
But this Manichæistic view has surely by this
time been discredited. It can be perceived that
no active outside opposition is necessary, none
beyond the unruly wills and affections of sinful
men. Matter has to be coerced : it is coerced
by every artist into the service and manifestation
of mind. There is real work in hewing a stone
out of a quarry and placing it in its proper place
in a cathedral. The energy expended is all
understood ; the forces operating are within our

control : they all come within the scope of physics
—all except the " plan " which determined the
right place, the particular niche in a consistent
scheme, of which the beauty and design shall
become apparent. No more energy was needed
to place the stone in its right position than in a
wrong one ; but the increase of value in the
stone, when truly adjusted as part of the finished
fabric, is something that does not belong to
physics, or to any branch of science : it belongs
to æsthetics.

The achievement of Beauty is not accounted
for by physics ; no, nor by biology. There is a
beauty before which we can only stand in venera-
tion : it is beyond explanation. To ask why a
thing is beautiful may be like asking why it
exists. Utility, preservation of the race, all
manner of explanations have been attempted ;
but all fail. There can hardly be anything
utilitarian about the beauty of a flower or a tree,
none at all about the beauty of a sunset. So far
as science is concerned, a flower would equally
serve its purpose if it were merely conspicuous.
Indeed, as regards the ultimate purpose of existence, and the reason of the struggle for existence, science has little or nothing to say. We
are constantly in a region where we have to work
by faith, not by sight ; by imagination, not by
knowledge. The value of existence is incal-
culable ; it cannot be reduced to a formula.

But if there is no active opposition, does that
mean that everything is easy ? By no means.

The engineer, in constructing a bridge or devising a steamship, need have no active opposition to encounter : he may find his fellows all ready to help ; and of course their help is needed.   What he has to overcome is mainly the inertia and refractory character of matter.   As I have pointed out elsewhere (in *Evolution and Creation*), inertia is a sufficient explanation of the need for force and energy.   The stone that has to be raised does not struggle against us, as a live thing might ; it is purely inert.   But to overcome the inertia, force is just as necessary as if instead of mere inertia there were an equivalent amount of active opposition.

For some reason or other, the Universe appears to be full of inertia.   The making of a world takes time ; it has to go through a long series of processes.   In its passage from a whirling mass of gas into a constellation of stars, with their surrounding solid planets on which life as we know it may gradually become possible, millions of ages must elapse.   So, at least, it has been with the only solar system that we really know ; and it seems blasphemous to suppose that it could have been achieved by any other or better plan than the one adopted.

If we are puzzled by the question why so much time is needed for the process of evolution, we should remember that the question of much or little time is not important : so long as some time is involved, it matters little how much. In every natural process time is involved, even in the alteration of the configuration of an atom, though to us that is infinitesimal. Nothing springs into existence ready-made ; in other words, everything grows. A fruit is not born suddenly, but has to go through the stages of bud and flower ; indeed, through many stages of tree-development before that. Stages seem inevitable in every natural process ; and so there have been stages in the history of this planet which to us seem almost infinitely long.

Duration, as Bergson says, seems an essential ingredient in everything. And intermediate stages must seem imperfect to those who have sufficient knowledge to be acquainted with the finished product. A builder's yard is not beautiful ; it is interesting, and is a stage towards a finished building. The same may be said of a fœtus— it is ugly. The same, I suggest, may be said of some present conditions and aspects of the human race. Ugliness is incompleteness, an unfinished stage in the process.

> . . . none but Gods could build this house of ours,
> So beautiful, vast, various, so beyond
> All work of man, yet, like all work of man,
> A beauty with defect—till That which knows,
> And is not known, but felt thro' what we feel
> Within ourselves is highest, shall descend
> On this half-deed, and shape it at the last
> According to the Highest in the Highest.

We need never be surprised at imperfection ; we can always regard it as a stage to something higher. If everything were already perfect, the universe would be dull, and there would be no growth or development. Such a universe as that evidently does not exist. Hence those who object to evolution are objecting to an essential feature in the Universe, and I am afraid must be stigmatised as stupid. This is not an adjective to throw about lightly, but it seems applicable to those who wish things to be hurriedly made other than they are. That is the folly of all red revolution, or what is now called Bolshevism. It is as if a gardener, wishing a tree of certain shape, could not wait for patient training, but set to work with axe and saw to smite it into shape ; thereby wasting the sap and ruining the tree—or the nation.

There must be some reason why time is necessary in every process ; though at present we may not know why, and though we may be even doubtful as to what " time " really is. Existence is evidently not limited to the slice we call the present. Before we can think of it, it has become the past. In what sense the past still exists may be an unanswerable question ; and whether the future in any sense exists, a still more unanswerable one. Whatever else time is, it must be the operation of turning the future into the past, passing instantaneously through the present on the way. In that sense time must be real.

> The splendours of the firmament of time
> May be eclipsed, but are extinguished not.
> —Shelley, *Adonaïs*.

The inorganic or inanimate world lives wholly in the present : it obeys every impulse ; it is without care, memory, or forethought ; it submits without a trace of rebellion, without a trace of co-operation, to anything that happens ; it lays by nothing for the future. But an animal does not live only in the present ; it remembers the past to some extent ; it anticipates and makes provision for the future. Surely it is by the past and the future together that we try to regulate and arrange our lives. In so far as there is anything that can be called " preparation " in the inorganic material world, it must be due to the management of something outside it. Any such preparation is evidence—even if not proof—of intention and design and effort.

When we contemplate the serious efforts in preparing our dwelling-place, the pains taken for our benefit and for the amelioration of our lot, and then think of the unnecessary troubles caused, say, by the lack of co-operation between capital and labour, and emphasised by industrial disputes,—when we think of the internecine warfare between nations, who might be able to help one another in the struggle against natural difficulties,—we are apt to be dismayed. But that is to take a short-sighted view. By faith we can realise that even that struggle has been or is somehow necessary ; that it contributes something to experience; that it is only a stage in the process. The problem of evil looms large and difficult, but may turn out to be not insoluble.

A knowledge of good and evil was a step up in human development and led to an invigorating conflict. Can evil be the shadow of good? Light and shade are essential to a picture ; if all were equally bright there would be nothing to see.

> No ill no good! such counter-terms, my son,
> Are border-races, holding, each its own
> By endless war.

The opposing forces need not be due to hostile Principalities and Powers, but may be accounted for, partly by the inertia of matter, mainly by the obstructiveness, stupidity, ignorance, and self-will of man.

That the earth can be reasonably said to have gone through a long labour of preparation before a human race could exist upon it, is an illuminating and instructive fact. That long period of gestation must increase our sense of responsibility ; and although we may be occasionally dismayed at the thought of how imperfect we are, and how unworthy of all the pains that have been bestowed upon us, yet we may be thankful that we have now at length become conscious of the effort. Unlike our lowly ancestry, we can feel that at last we have become partners and co-operators in a process which, after all, is only in the beginning ; and which science has begun to tell us may last without intermission and, still under beneficent conditions receiving the heat and light of a sun in a stable solar system, almost interminably. For though the Sun must contain seeds of its own material decay and ultimate end as a luminous vivifying power, yet it seems likely to remain brilliant for millions of years, and it may be of centuries.

What science has to say on the subject of gradual evolution and adaptation to conditions, it says with no uncertain voice. Even those who exclude the Deity from their consideration, and have " no need of that hypothesis," can see that everything that has survived is well adapted to the conditions. The more they penetrate the secrets of nature, the more do scientific men realise the wealth of existence, and the beauty of adaptation that has made it possible. This applies not to live things alone, but to everything. The structure of an atom is like the structure of a solar system : every piece of matter is a sort of organism, of which the parts are as well adapted to each other as the protoplasmic constituents of our own bodies.

Professor Whitehead, that philosophic mathematician, has said that " science is taking on a new aspect which is neither purely physical nor purely biological. It is becoming the study of organisms. Biology is the study of the larger organisms ; whereas physics is the study of the smaller organisms." He here adopts the view that I have been emphasising, that systematic organisation is the rule in the physical universe. The organisms studied by biology are more complex than those studied by physics and chemistry ; they involve an entelechy or guiding principle that lies outside our physical and chemical scope ; but physics and chemistry deal with organisms too. An organisation of electrons and protons, those two apparently simple and fundamental units, aggregates first into atoms, then into molecules, then chemical compounds, then crystals, and so into all the things which we can see and handle and dissect. These also appear to be in some way the product of an evolutionary and gradual process—a process apparently of falling temperature, down to a level at which complex atoms could form and persist without breaking up into their constituent units. Only on cool planets could many of our atoms form. The stars are too hot ; the matter in them is in an elementary stage of evolution.

At some further stage in the process, long after the formation of the heavier atoms and molecules, when the chemical constitution is sufficiently complex—and when the material is suitable for incorporation into cells,—first life, and then mind make their appearance ; one in all probability being the rudiment of the other.

Problems Connected with the Physical Basis of Life.

Life is studied in the science of Biology, a most comprehensive science, including not only Natural History, but Anatomy and Physiology as well. From the Presidential Address entitled

"Function and Design," by Professor J. B. Leathes, F.R.S., to the Physiological Section of the British Association Meeting at Oxford in 1926, I make the following extract. It discriminates between the biological sciences, and expresses the relations between life and matter in an expert and instructive way :

Among natural sciences physiology takes a place which in one respect is different from that taken by any other. It studies the phenomena of life, but more particularly the ways in which these phenomena are related to the maintenance of life. Anatomy and morphology are concerned with the forms of living organisms and their structure ; biological chemistry, as distinct from physiology, with the composition of the material in which the phenomena of life are exhibited. The province of physiology is to ascertain the contributions that certain forms and materials make to the organisation of the living mechanism, and learn how they minister to the maintenance of its life. Function implies ministration ; structure for physiology implies adaptation to function, what, in a word, may be termed design.

Ultimate analysis of the phenomena with which physiology deals leads to a fundamental distinction between matter in which life is manifested and matter in which it is not. Life is exhibited only in aqueous systems containing unstable perishable combinations of carbon with hydrogen, nitrogen, sulphur, phosphorus, and oxygen, in the presence of certain inorganic ions which are present in the sea—the native environment originally of all forms of life. The inalienable property that such matter exhibits when alive, and that matter which is not alive does not, is that these unstable organic combinations are for ever re-forming themselves out of simpler combinations that do not exhibit this property.

The address goes on to say that of the various chemical components of protoplasm (which is the name for the basic material of living matter),

proteins are generally considered the most important. . . .
The best-known varieties of proteins . . . consist of chains of about a hundred, sometimes nearly two hundred, links.

Each link is a chemical compound known as an amino acid, of which there are many varieties ; and the links might be all different.

In any such isolated protein it is probable that the order as well as the proportion in which each amino acid occurs in the molecule is fixed, and it is this specific order and proportion that accounts for the specific character and properties of the protein.

That is to say, the order and arrangements and constituents of the molecule are supposed to determine its specialised function ; for it is manifest that the different cells in the body have different functions to perform.   The protoplasm in the kidney differs from that in the heart or the brain : again, it has a special function in the retina of the eye ; while its functions seem reduced to a minimum in hair and nails.   The way that each particle of food goes to its own place, and there functions as it ought, is surely puzzling.   Nevertheless that is what happens. And the possible permutations and combinations among the constituent molecules even of a single protein are numerous enough to account for any variety of function—provided that that is the right way of accounting for it.

Thus, for instance, if the chain above spoken of contained only fifty links, and if there were only nineteen different varieties in it, Professor Leathes calculates, in accordance with straightforward rules of arithmetic, that the number of different arrangements of its parts could be 10^{48}. Now this is an enormous number, about equal to the number of atoms in a shell of the whole earth one mile deep ; a number of atoms which, if placed in a row, would reach right across the diameter of the Milky Way and back, more than a million-million times. This multiplicity of units in the cell, then, is among the considerations which confront a biologist when he is dealing with the material basis of life.

But Professor Leathes goes on to say that in the chemical make-up of protoplasm, proteins, though the most abundant component, are not the only ones that are necessary. Pre-eminent among the others are what are called the nucleic acids, which apparently form 40 per cent. of the cells

into which are packed from the beginning all that pre-ordains, if not our fate and fortunes, at least our bodily characteristics down to the colour of our eyelashes ; (so that) it becomes a question whether the virtues of nucleic acids may not rival those of amino acid chains in their vital importance. From Steudel's figures it can be reckoned that there are about half a million molecules of nucleic acid in a single sperm-cell of the species with which he was working.

If biologists sometimes seem to lose their way among the stupendous phenomena of life, and the processes which occur in living matter, we need not be surprised. The constitution of the heavenly bodies and the structure of the atom are in a sense elementary, compared with the problems which confront those who would ascertain and disentangle even the material aspect presented by living things. And how much more must be accomplished before we can hope to understand the material basis of mind. Sometimes it seems as if science were only scratching the surface of the Universe. Science is, after all, only a few centuries old; and the progress before it is infinite.

I take this as an example of only one of the infinite stages of progress, that, as we are now beginning to see, lie open to humanity. The scientific aspect of human progress is only one of many; but the outlook even in that direction is of overwhelming magnitude.

PROBLEMS OF LIFE AND MIND.

Whence life and mind come we know not, nor what is their nature. But here they are, and it is madness to deny them. Some, indeed, of those who have explored the possibilities of the higher organisms, and their mental and psychical activities, have discovered evidence that they are related to another order of things : that the Universe is not limited to the physical and the chemical ; but that amid all the possibilities of existence the spiritual, too, is a reality, and that through its manifestations by physical means we are beginning to get into touch, even scientific and demonstrational touch, with a higher order of being, whose existence has hitherto been postulated only by Religion.

For already humanity feels that in its essence it is in closer connexion with these unseen things than with the materials among which it lives, and amid which it finds its difficulties. We *are* associated with the physical world for a time, but we do not really belong to it. Our innermost nature belongs to a spiritual world, which, I would say, is certainly with us all the time, though we have no sense-organ for its appreciation. Theologians, doubtless, have always realised that : Saints and Seers and Prophets have emphasised it continually. It would be wise to listen to those with greater insight than ourselves.

PROBLEMS OF LIFE AND DEATH AND THE FUTURE.

To the thinking and contemplative mind the Universe is full of problems ; most of them insoluble. Our gradual growth in scientific knowledge rather intensifies than removes them. Science solves some problems, but always discovers others : it pushes back the boundary of ignorance, and the result is to enlarge our contact with the unknown. Science starts with certain postulates, such as the existence of matter and energy, and seeks to develop the laws of their interaction : it does not attempt to account for

them.   It accepts the Universe as a going con-
cern, and seeks to ascertain the laws according to
which it works.   Ultimate problems it leaves to
Philosophy ; or else, perhaps rather contemptu-
ously, to Religion.   Religion does not attempt to
solve them in any scientific manner ; it soars
above them in an atmosphere of faith.   The
problems that chiefly perplex the average man
are not the ultimate problems of existence, but
the daily experiences, the troubles and trials and
temptations, around him.   These are perforce
brought to his notice ; they often lead to great
distress, and from them he may try to escape by
ignoring them and turning his attention to what
seem to him positive concrete realities—the
obvious business of this world.   At lax times he
may go so far as to comfort himself with the hope
that the people who have found refuge in religion
may be right ; that the safest course is to assume
their general rightness, and to hope for the best.

If we were asked what are the problems which
loom largest in the mind of the ordinary human
being, we might answer, surely those which centre
round the uncertainties and contingencies of life,
especially those connected with the lamentable
but undoubted facts of death and bereavement.
The consolations of religion mitigate these for
some people, but by no means for all.   Nor does
it appear to be the custom to deal with them in a
whole-hearted and satisfactory manner.   We are
all apt to be brought up against a problem of
bereavement, and feel that we have nothing to

say. Some human beings, especially the old,
live in constant fear of death, which they know to
be inevitable, which they sometimes try not to
mention or even think about, but which forms a
gloomy background to their lives.

The Burial Service is not exhilarating ; the
disposal of the corpse is a repulsive necessity.

> I hate the black negation of the bier.

Outlook on life is overshadowed by these things,

> My son, the world is dark with griefs and graves,
> So dark that men cry out against the Heavens.
> Who knows but that the darkness is in man ?

General Wolfe, before taking Quebec, read
with admiration Gray's *Elegy*, containing the line:

> The paths of glory lead but to the grave.

One of the odes of Horace, too (I. 28), expresses
the melancholy outcome of all human learning
and grandeur.   And the pessimistic feeling that,
after all, human effort is futile and progress
illusory, is probably a mood more prevalent than
is generally admitted.

> The plowman passes, bent with pain,
>   To mix with what he plow'd ;
> The poet whom his Age would quote
>   As heir of endless fame—
> He knows not ev'n the book he wrote,
>   Not even his own name.

(This and several other quotations are from Tennyson s
The Ancient Sage, though this last is not one of the Sage's
own utterances.)

Some take refuge in Philosophy ; more in Religion ; but the majority are unacquainted with either ; they are content with a blind instinct for material well-being, and live in the present, resisting any effort to look beyond. The life of animals may be supposed happy, because they do not look before and after. But, for better or for worse, man has attained the power of brooding on the past and anticipating the future ; and some anticipate the future with alarm. It cannot be said that Religion has widely removed the element of fear. Is it not true that, in some not wholly extinct forms, it has rather intensified fear, by insisting on the hereditary quality of sin and by the doctrine of eternal punishment? To some the doctrine of vicarious propitiation has brought comfort ; but to many more that idea seems illusory, and perhaps even unfair. Fear is a terrible bugbear, from which faith and knowledge should free us. The fear of God is the beginning of wisdom, we are told ; the word translated " fear " may probably mean " reverence," but, even so, it is but a beginning. The love of God is a more helpful, though truly a more difficult and more advanced thought. Fear of the unknown is natural to savages, and is responsible for much superstition. Fear of the unknown has not yet been eradicated and replaced by eager anticipation and vivid hope. That seems at present too much to expect. Is it possible that the science of the future may rush in and restore confidence and bring consolation where religion has failed ?   Certainly materialistic philosophy will not succeed ; but the tendency at the present day is for science in its philosophic mood to become less materialistic, to realise that matter is only something which conspicuously affects our senses, that we do not see things as they really are, that matter itself may be resolvable into something else, and that even the material universe may ultimately be found to have an idealistic constitution, and to be suffused throughout with spiritual reality.

Scientific men, too, are beginning to apprehend something of the truth, and even to forecast a unification.   One of the great astronomers of all time, who is now with us in the Newtonian University of Cambridge, Professor Eddington, has expressed his own conviction thus :

> I venture to say that the division of the external world into a material world and a spiritual world is superficial ;

and he goes on to say that if there is a line of cleavage, if there is ultimately a real breach of continuity (which for myself I think unlikely), the cleavage is something other than that.

PROBLEMS OF INDIVIDUALITY OR PERSONALITY AND ITS PERMANENCE.

Already we are beginning to find out that mind can act on mind in other ways than through the organs of sense.   Our sense organs, which

K

we inherit from our animal ancestry, are beautiful structures ; but they are very limited in their scope ; and our minds have already far transcended their direct indications. What senses and powers we may hereafter possess we do not know. This planetary episode is of great importance as a means for forming character and developing individuality ; but, once developed, the reality is not dependent on the means that has helped to produce it. The material body has had its day, and may cease to be ; but existence does not cease. Even the material body does not cease ; its atoms continue, though they are then, like a crowd or any other assemblage, dispersed and scattered, ready to form another and another frame of things for ever.

It may be said that some of what I uphold is speculation ; let us grant that it is : but it is speculation, like other scientific inferences, based upon scrutiny of facts. The facts are not all generally admitted : further study is necessary to confirm and strengthen conviction in their truth, but those who have most studied the matter are most convinced that there are facts not yet recognised by orthodox science. On what the meaning of these facts may be, there is room for legitimate difference of opinion ; but until those facts are taken into account it is unwise to deny that they may have a deep and abiding influence on human progress and human happiness. John Masefield urges us to have faith in what makes for ultimate and real happiness :

. . . trust the happy moments—what they gave
Makes man less fearful of the certain grave
And gives his work compassion and new eyes :
The days that make us happy make us wise.

The question whether there is individual survival of bodily death is surely a theme for scientific
inquiry.   It may not be answerable, but it can
be asked ;   and any question which can be
reasonably asked may be competent to receive
an answer some day.   An answer may in due
time be forthcoming if we patiently pursue, without haste, without rest, the ordinary procedure
of investigation which has already carried us so
far.   Philosophy and Religion may and will have
much more to say, in regions which science cannot
attain to, but this question is a simple one ; and
this question many believe that science is already
in the act of answering.

Who knows ? or whether this earth-narrow life
Be yet but yolk, and forming in the shell ?

None of us is able to predict what the science
of the future will be ; but many feel assured that
it will not be limited to a study of the forms taken
by organised bodies.   The Universe contains
more than that.   Science can advance in the
psychical direction as well as in the physical.
Indeed, it may ultimately find that there is no
sharp line of demarcation, no artificial boundary ;
that the two are interconnected ; and that, as
mind already dominates and controls and guides
and plans and arranges, so the spiritual aspect of

the Universe will be found to be dominant—
controlling, guiding, and directing the material
aspect in ways of which at present we have only
the dimmest conception.   The central doctrine
of that great seer and poet Virgil is suggested by
the familiar lines from the Sixth Book of the
Æneid :

> Spiritus intus alit, totamque infusa per artus
> Mens agitat molem, et magno se corpore miscet.[1]

The truths of Religion, in so far as they are
truths, can be probed and realised more vividly
than ever before.   The foundation or substratum of all religion is the recognition of a
spiritual world, and a sense of the real existence
and activity of Intelligences far above us in the
scale of existence.   These existences have hitherto
only been apprehended by faith.   The imagination of poets has soared among them, the intuition
of saints has seized upon them as the ultimate
realities ; so that this present life, with all that it
contains, its strange mixture of joy and pain, has
in some moods been felt to be as nothing compared with the glories that shall be revealed.

> I am borne darkly, fearfully, afar !
> Whilst burning through the inmost veil of heaven,
> The soul of Adonaïs, like a star,
> Beacons from the abode where the Eternal are.

Hail, then, to our destiny !   Our course is
illumined, as Plato also said, by a guiding light
from those who have gone before.

[1] Spirit sustains everything ; and, permeating every part, mind
controls matter and mingles with its mighty frame.

The Universe is far greater than we know ; the possibilities of existence are illimitable. That which we apprehend is a very small portion of Reality. The Universe is infinite in an infinite number of ways. It is no time to lose heart and feel that we are groping in the dark. The mists are being lifted, a light is shining through : towers and pinnacles of beauty are visible at times, although too soon the clouds of earth gather round them again. As says Francis Thompson :

> Yet ever and anon a trumpet sounds
> From the hid battlements of Eternity ;
> Those shaken mists a space unsettle, then
> Round the half-glimpsed turrets slowly wash again.

Our physical eyes tell us only a little ; our minds already reach far beyond their scope. With physical eyes alone we cannot penetrate the depths of reality. We already know much of which we have no sensuous cognizance ; and in that direction, both as individuals and as a race, we may look forward to advance towards heights unspeakable. An inspired writer has truly said that now we see as through a glass, darkly ; but then face to face. Now we know in part, but then, in the process of time—it may be ages hence—we shall know even as also we are known.

DEATH AND HEREAFTER

THE PROBLEM OF SURVIVAL

In speaking or thinking about human progress we are apt to be brought up against one of the problems that we began to deal with last time, namely, the question, What is the good of it all, if, as many scientific men have thought, it ends in extinction ? It cannot really be true, in a prosaic and literal sense, that the paths of glory lead but to the grave ; yet in certain moods that seems to be the dispiriting conclusion. So also it has been felt, now and again, that the path of effort, labour, sacrifice, all the learning and achievements of man, his aspirations and heroic struggles, can ultimately have but one end, the death of the individual very soon, the death of the planet in due time. And just as the turmoil of a city goes on serenely after we are gone, and no matter who has perished, so the Universe will continue placidly on its way long after the sun has faded and the earth become deadly cold, and all the toils and aspirations of humanity have gone into nothingness, as if they had never been.

Whether cold or heat will mark the end of the earth we know not : we know that in the course of ages it must come to an end somehow. There is a remote possibility that it might end in a fiery outburst, if the sun in its agelong course through space encountered some region fuller of matter than usual, a misty nebula, the friction of which would quickly raise everything to a white heat.

There was a time when some such catastrophe —"the elements shall melt with fervent heat"— seemed not very far distant. Blazes or conflagrations do occur at times in various parts of the heavens. Why should the sun be a star exempt from such a contingency? There was a time also when it was calculated that the sun's source of heat must at a somewhat distant epoch be exhausted, and no longer able to support life on its planets as it does now. Modern science has extended the probable lifetime of the solar system enormously. A new source of heat has been discovered in the energy of the atoms ; so that the fading of the sun is no longer a reasonable contingency to be expected for millions of centuries—and in such a period as that almost anything may have happened, and humanity may have risen to heights as yet inconceivable.

Nevertheless, it remains true that sooner or later there must be an end, even of the solar system. In its present state it did not always exist : neither can it always continue. While as to the individual, if we limit ourselves to his bodily organism, the end is close at hand. Literature is full of the thoughts thus engendered. The theme of Omar Khayyám is an urge to catch the fleeting moment, to make the most of it while it lasts.

> The Stars are setting and the Caravan
> Starts for the Dawn of nothing.—Oh, make haste !

The *Archytas* ode of Horace represents the end of human glory and learning, depicting it as a corpse on the seashore silently beseeching a passing stranger to cover it with some handfuls of sand.   Or, as F. W. H. Myers once sang, having evidently this ode of Horace (I. 28) in mind, in a poem the title of which should be " Mortality "—a sort of dirge on the end of the human era :

> Lo all that age is as a speck of sand
> Lost on the long beach when the tides are free,
> And no man metes it in his hollow hand
> Nor cares to ponder it, how small it be ;
> At ebb it lies forgotten on the land
> And at full tide forgotten in the sea.

Pessimistic moods are apt to occur to every-one from time to time.   Bereavement is a terrible thing, and has led many to despair.   The wonder is that such thoughts, if they are persistent and part of our accepted philosophy, do not lead to a cessation of effort, a throwing up of the hands, an abandonment of hope.   Fortunately we are not guided only by what may be a mistaken philosophy : we are guided in our actions mainly by instinct ; and even those who think that this life is all, still work and strive, and often sacrifice themselves for others, and generally act in a way which, though hardly logically consistent with their philosophy, we yet admire and recognise as among the higher attributes of man.   The individual is often superior to his faith.

But a question inevitably rises, Is the philosophy true ?   Is this kind of pessimism justified by the facts ?   Those who hold, with one phase of present day orthodox science, that personality is a bundle of reflexes, that memory is lodged in the brain, that ideas, aspirations, thoughts, are all inextricably involved in that organ, and cannot exist without it—those, in fact, who hold that the body is the whole man, that he is simply a piece of mechanism which must wear out and run down, that "vitality " as displayed by an organism has no more meaning than the " horology " of a clock—can hardly resist a pessimistic conclusion if they press their beliefs to a logical outcome. They may, nevertheless, say that life is worth while, that human relationships are of value, that our existence has all the more value while it lasts, because it is so soon coming to an end ; that now is our only opportunity for service, and that we must make the most of it.   Or, as the great mathematician, W. K. Clifford, who had no faith in survival, expressed it in his essay called " The First and Last Catastrophe " :—" Do I seem to say : Let us eat and drink, for to-morrow we die ?   Far from it ; on the contrary I say : Let us take hands and help, for this day we are alive together."

But still the reality of death must occasionally press hard upon those with a materialistic outlook. They know, we all know, that the existence of the brain is very temporary, that it will be burnt

or buried, that it will disintegrate into its constituent atoms : and any permanent destiny for
the mind is relegated at present by an energetic
and dominant fraction of orthodox science to
the region of superstition. You know well that
scientific discoveries, though they have greatly
extended our estimate of the life of the planet, and
therefore presumably of the human race, have
been unable to extend the life of the individual by
more than a few years. What used to be said
about a man of seventy is now more usually said
about a man of ninety, and presently it may be
extended to a hundred. But what, after all, is a
century in the illimitable æons of existence !

Sooner or later science will find itself bound
seriously to ask, and attempt to answer, the
question, Is death the end ? Is there any
meaning in human survival ? Appearances are
against it ; death does look like the end ; but we
must not be misled by superficial appearances.
May it not be really true, in spite of appearance,
that character, memory, affection, are in some
way *in*dependent of the brain, and can continue
without it ?   Is the brain, after all, not the mind,
but only an instrument for the manifestation of
mind, just a sort of trigger for operating the
muscles, and thus acting on the material world ?
Is it a machine, not for planning, but for getting
planned things done on the surface of this planet
during the short time that the instrument lasts ?
And is the experience thus gained, the learning
acquired, the character formed, the personality,

a persistent entity which can continue amid other surroundings, quite apart from the matter in which we recognise it as incarnate here and now ? These are questions that science is beginning to ask ; and I believe it will find—some of us say has found—answers in the affirmative.

Lord Kelvin, in an inspired mood, said that by the everlasting laws of honour Science was bound to face fearlessly any problem that could reasonably be presented to it. Sooner or later it must face the problem of survival. Some scientific men are beginning to face it now. Evidence clearly bearing on this point is known to a multitude of people. Facts are accumulating, and have accumulated, which must sooner or later be incorporated into science. The deaf ear and the blind eye cannot always be turned to these things. Continuity and persistence of existence, at least of the fundamental material elements, are already admitted and enforced by science ; continuity and persistence of existence are not yet scientifically asserted of the human mind. While as to the human spirit—the term has not even been defined.

Meanwhile the word " incarnation " is abhorrent to the materialistic scientist. Nevertheless, it may connote a momentous fact. If what I hold is true, the body is a serviceable subordinate to the spirit of man. The spirit, and all that it involves, is a permanent entity, which makes use of the properties of matter, not as essential to its existence, but as instrumental aids essential to demonstration.   The material body may be—like the pen of a poet, or the violin of a musician— an instrument without which he cannot communicate his ideas to his fellows while they, too, are incarnate in the bodily mechanism ; for that mechanism bounds and limits and screens them from their real existence, of which through their bodily senses they can catch no glimpse.

To me it seems that science is in the act of adducing evidence for, and gradually proving, the existence of a spiritual world ;—that is to say, the existence of beings and intelligences in grades as numerous above present-day humanity as we already know them to be below—and that thus our whole outlook on existence will be changed, and our data immensely enlarged.   We are finding that memory, affection, character, and personality are not a part of, and are not limited by, the aggregation of atoms with which we move about on the planet, but are something far more real and permanent, more deeply-rooted in the nature of things.   The discarding of the body may be a painful episode, but not more painful than many illnesses ; often not consciously painful at all.   We shall gradually come to look upon death as an adventure, a sort of enterprise, which we can undertake with hope and faith, and with a sort of joy in the prospects of the wider interests upon which we are entering when separated from the matter body.   For, after all, this body, in spite of its physiological perfection, is troublesome at times.

It binds us down in space and time ; it is a limiting and restricting thing, like luggage ; and liberation from it may be an emancipation.

Those who have nearly gone through the process, and have come back, tell us that that has been their experience, and that it was the coming back that was chiefly painful. While those few of us who feel that they have been privileged to have communication with their friends on the other side have had encouraging information given them about the conditions, and rejoice in a sense of communion. They have thereby emancipated themselves from terror of the unknown, have freed themselves from undue lamentation over bereavement, and have escaped from dread of the future. They have learnt that no acquisition will be lost or wasted ; that opportunities for progress will continue ; that all true powers and aptitudes persist, and will gradually become enhanced. They learn also that individuals will still remain themselves ; that they will be recognised by, and will recognise, their friends ; and that they will all still be in an active world of love and service.

Furthermore, it turns out that even our appearance will not be greatly changed, that we shall still have a bodily mode of manifestation, that we shall not be out of touch with the physical world, although the particles of matter which we have accreted from the earth will no longer belong to us. Our identity never depended on the identity of those particles, neither will it cease when they are completely left behind. The spiritual and the physical will still be interlocked ; a bodily mode of manifestation suited to the new surroundings need not be composed of atoms of matter.

My own surmise is that our true bodies are not made of matter now ; that the action of mind on matter is not direct, but always occurs through the intermediation of something whose properties we try to summarise under the name of the Ether of Space.   What was called " the spiritual body " is asserted to be an etheric body ; not an unsubstantial evanescent reality, but an entity with properties far more perfect than any form of molecular matter.   Indeed, we are beginning to learn that the atoms themselves are interconnected through this same medium now ; that it holds them together and gives them their configuration ; that through it the atoms of terrestrial matter have been built up into our own visible bodies, including those very sense organs through which we apprehend the material aspect of things— organs that are indeed effectively stimulated by nothing else. (See *Ether and Reality*.)   Hereafter we may find ourselves more directly associated with the ether, and with matter hardly at all.   The ether probably has many other modes of activity, unapprehended by our present senses, and those etheric properties and powers may be what we shall hereafter utilise.   I believe we are even now acting on the physical world through the ether, that we have no other mode of acting on

it, that our action on matter is indirect. Matter
is discontinuous, so that even when matter acts
on matter it always does so through the ether ;
a link, a connecting medium, is essential. When
mind acts on matter it must be through the
ether too. That is what we really inhabit ; that is
already, and will always be, our spiritual home.

I already feel in my bones that these things are
true, though I am willing to admit that scientifically speaking the demonstration is incomplete or
lacking. Suppose, however, that some day the
demonstration is admitted, the fact ascertained ;
that whether with the body or without the body
existence is continuous. Surely that knowledge
would have an immense effect upon our outlook,
perhaps even on our conduct. Every truth, no
matter how small, has a great influence. There
are no half-truths : if a thing is true it is completely true, and its consequences may be infinite.
Once a truth is realised, we find that it has always
been there : it is our recognition of it that is new.
Electrons, X-rays, all the multitude of recent
discoveries have been in existence all the time ;
only we did not know.

So it will turn out with this question of spiritual
existence and survival. We are in process of
discovering a whole new world, nothing less.
Some day the discovery will be admitted, which
many individuals have begun to make now.
And then the corporate outlook of mankind on
the Universe will be changed enormously. Life
will no longer be a frustrated episode, a flash in

the pan, temporary apparition, coming out of nonentity, moving about with pomp and circumstance, and then returning into nonentity. Death will no longer be regarded with gloomy apprehension, and dreaded more and more the nearer we approach it. Nor need loving friends watch by our bedside with anxious despair and dread " the surly sullen bell." Shakespeare likens impending bereavement to the period of autumn :

> Bare ruin'd quiers, where late the sweet birds sang.

He also likens it to the colours of the sunset :

> Which by and bye black night doth take away.

And although the sense of impending loss

> . . . makes thy love more strong,
> To love that well, which thou must leave ere long,

he comforts his friend, in Sonnet 74, thus :

> But be contented : when that fell arrest,
> Without all bail shall carry me away,
> My life hath in this line some interest,
> Which for memorial still with thee shall stay.
> When thou reviewest this, thou dost review
> The very part was consecrate to thee,
> The earth can have but earth, which is his due,
> My spirit is thine, the better part of me :
> So, then, thou hast but lost the dregs of life,
> The prey of worms, my body being dead,
> The coward conquest of a wretch's knife,
> Too base of thee to be remembered.
> > The worth of that, is that which it contains,
> > And that is this, and this with thee remains.

Let us now cease to generalise, cease to anticipate the future discoveries of science. Let us
try to state in a quite simple manner the kind of
way in which those who are convinced of the
reality of survival are already beginning to
contemplate death.

" Death " need not be regarded as a gloomy
subject ; there is no need for it to be so considered ;
it is rather like emigration. There is something
sorrowful even about emigration. When a youth
sets out for a distant country he may not be sad,
but his relatives are ; and yet they may have
hopes that it will be the beginning of a bright and
useful and prosperous career. It is of the nature
of an adventure : it is not by any means all
gloom. There is the pathos of parting, but
there is hope and joy also.

Now, to understand what death is, we must
first know what life is. I have written elsewhere
on this subject, in *Life and Matter*, and must be
brief. Life is a very difficult thing to define :
we know something about it, however. We
know that it is not a form of energy, but that it
is a guiding and directing principle. It uses
energy and it uses matter, but it does not seem
itself to belong to the physical frame of things
at all. Without a supply of physical energy life
would be helpless and inoperative, but on this
planet, and presumably on all inhabited planets,
a supply of physical energy is available through
purely inorganic and molecular processes, which
go on independently. And yet somehow they

can be interfered with and modified, as if by something outside; something controlling not indeed
the movements of the atoms but the arrangements
of the molecules.   Life seems able to produce a
grouping of molecules that is essentially a re-
arrangement of atoms, such as would not have
occurred without its influence.

How it does that we do not know : we do not
know what life is.   It is easier to say what life
is not.   And, first of all, though it is tempting to
say that life is energy, yet when we use the term
energy in the technical scientific sense, as a
physicist uses it, the assertion that life is energy
is not true.   It is perhaps rash to make that
assertion ; for twenty or thirty years ago we
should have said, or most of us would have been
inclined to say, that matter was not energy.   The
motion of matter is energy, just as a current of
electricity is energy ; any forced arrangement of
matter or of electricity has energetic properties ;
but that does not mean that either matter or
electricity is energy.   Given certain conditions,
energy can always change its form ; it can be
transferred from one body to another, and can
thereby do work.   Energy is something that can
be stored and liberated ; and when liberated work
is done and some effect is produced.   The energy
is not consumed, but it is transferred and transformed.   It can exist at one moment in a state of
strain, like a bent bow, or a coiled-up spring, or a
raised weight, or an unstable chemical compound,
in an apparently quiescent form ; but in the next

instant it can be liberated and display itself as motion, the same in amount as before, but in a different form. The characteristic of energy is that it is constant in amount, but protean in form ; and all activity is due to the passage of energy from one form to another. Energy is conserved, and the conservation of energy is one of the sheet-anchors of science.

Now life does not appear to be conserved. It seems to be abundant in amount, without limit ; coming we know not whence, and departing we know not whither. It interacts with matter for a time, and we only know of it when it is so interacting. Organic life can only exist when there is a supply of inorganic physical energy. It does not produce that energy; it makes use of it. Without such a supply it would seem to be helpless. We are living in a blaze of energy derived from the sun : all the activity of the earth —not only the growth of plants and animals, but the rivers and the winds—is due to ether vibrations, which come to us across space. That activity we have the power of directing : we can liberate it, or store it, or deal with it in various ways. Live things have the power of controlling but not of producing energy. That is why we say that life is a guiding and directing principle.

That life is not energy, in the technical sense, is seen by the fact that a seed, for instance, can give rise to countless generations : there is nothing conserved about it, no limiting quantity. One acorn could grow an oak, and thereby

thousands of acorns, and each one of these a fresh
tree, and this a multitude of acorns, and so on.
It is like the influx of something from outside, as
if we were tapping an infinite reservoir ; as if life
could by proper arrangement be brought to interact with matter for a time, and then left to depart
whence it came.   It is like something condens-
ing on the planet, utilising matter and energy,
and then evaporating away.   Meanwhile life
controls matter, and in our own case enables us
not only to plan and will and design, but to
execute and construct.   Under the influence of
life, things are produced which otherwise would
not have occurred, from a seashell to a cathedral.
The products of life vary immensely, from a
blade of grass to an oak.   Life produces special
phenomena, from a firefly to an electric arc, from
the song of a cricket to an oratorio.   All these
things are due to life in its interaction with
matter.   Life does not exert force, and yet it
contrives that force shall be exerted, and things
are moved.   It does not propel a train, for
instance ; it does not even directly guide it : a
train is propelled by a locomotive, and guided
by rails, but the general manager determines when
it shall start and where it shall go.   That is the
way life acts : it drives a train like the organiser,
the general manager, not like a locomotive.

But, before passing on, perhaps I ought to put
in a caution.   Less than fifty years ago we should
have denied that matter was a form of energy :
we should have said that matter without energy

was inert and helpless, but that supplied with
energy it became active.   We should have said
that matter was another thing that was conserved,
constant in quantity, merely altering its configuration, changing its form, but continuing unchanged
in amount.   We knew that matter could not be
destroyed by any of our operations.   The conservation of matter was the sheet-anchor of the
chemist : destruction of matter was impossible.
Whatever was done to it, into whatever combinations it entered, whether it was dissolved or
evaporated or burnt, it continued unchangeable
in amount : the chemical balance showed that it
weighed the same as before.   And indeed on
the earth that continues true ; but in the sun
and stars it now appears not to be true.   There
matter is not conserved : a radiating body is
radiating at the expense of part of its own substance ; in other words, matter is changing its
form and passing into energy.   The conservation
of matter, in any ultimate sense, has to be given
up.   Energy can increase in quantity at the
expense of matter.

Have we, then, to give up the conservation of
energy likewise ?   That does not follow.   I said
that the characteristic of energy was that it
changed its form.   Matter may be one of its
forms : and, if so, the conservation of energy
may still hold.   When I say that energy increases
at the expense of matter, I merely mean that it
is changing from the form of matter into the
form of radiation ; it is being transferred to

impalpable ether.   In that transference and trans-
formation it does work—it produces effects—in
accordance with the general law.   To retain the
conservation of energy we must assume that
matter is one of its forms.

This conception, that matter is a form of
energy, is rendered easier by the discovery that
matter is electrically constituted ; that it consists
of nothing but groupings of protons and electrons,
which in themselves are electric charges.   But
then the question is inevitably forced upon us,
What are those electric charges ?   Are they
forms of energy too ?   Can we resolve them into
intrinsic motion of something ?   Here we are
beginning to get out of our depth, and have
reached the confines of present knowledge ;
but sooner or later we are bound to take the
Ether into account, and some day we may find
that an electron is a whirling motion in the ether :
in which case the whole of the material world will
be simplified and reduced to one fundamental
substance in various states of motion.   The ideal
in physics, by no means yet achieved, is to reduce
everything to Ether and Motion.   This has been
the surmise of geniuses at different times throughout the latter half of the nineteenth century ; and
now in the twentieth century the idea is gaining
strength, and probably pointing the way to future
fundamental discoveries.

We must not pursue it further ; but what I
have said emphasises the caution which we should
feel about making rash assertions and saying that

life is not a form of energy. We know that
radiation is a form of energy, a form residing
wholly in the ether. We are now beginning to
regard matter as a form of energy, likewise
residing in the ether. And my caution is merely
this, not to be too sure that life may not be something residing in the ether too. We know too
little about it to be able to make positive assertions
one way or the other. We have far too much
still to learn. All our assertions should be
understood as limited by our present knowledge ;
the whole body of science is inevitably a summary
of our present knowledge about things. The
discovery that matter is a form of energy is a
revolutionary discovery, perhaps even yet not
fully made, but in the act of being made : and we
must keep our eyes open for other fundamental
and revolutionary discoveries. We may some
day find that the ether has not only physical
properties and attributes, it may conceivably have
psychic attributes also, and be related to life and
mind in a way of which we at present have no
conception. It may be that every psychical entity
must have some physical concomitant. Con-
versely, it has been suggested—though as I think
less plausibly—that every physical entity must
have a psychical concomitant. If invariable connexion should ever be established, it would be
a step toward ultimate unification. But that is
the task of Philosophy.

Meanwhile we must, as usual, be guided by
experience, and not shut our eyes to any of the

facts because we do not understand them. The main function of living things in the material world is to move and arrange matter. That is what we accomplish through our muscles, and that is all that we can physically accomplish. All the other operations that we see going on are conducted by the powers latent in the things themselves. We cannot build a crystal, but we can arrange that a crystal shall be built. We cannot grow a plant or a seed, but we can give a seed a chance of growing. We cannot construct a bird, but we can apply warmth so as to hatch out an egg. All that a gardener does is to move things ; the rest is accomplished by the mysterious processes of nature. Our own bodies are the result of those unconscious processes. As was said poetically long ago, we cannot by taking thought add one cubit to our stature.

No, we must, as I have just said, be guided by experience. We know that we ourselves can not only move and arrange matter, we can do something not apparently belonging to the physical universe at all ; we can plan and contrive and hope and love. We have a number of psychic attributes which do not belong to the physical frame of things ; and those, it may be, are our primary and persistent attributes. Our inter-action with matter seems a secondary and temporary thing. The puzzle rather lies, not in those primary attributes, which constitute our real self, but in the fact that we can interact with matter at all. To produce effects in the

world of matter we require something supplementary.   In order that life may interact with
the physical frame it is provided with a body.
We know what we mean by a body.   We
mean a mode of manifestation, an instrument.
A musician may have music in his soul, but he
requires an instrument to display it to others.
As the violin is to the musician, so is our body
to our soul.   We have constructed the body,
unconsciously, it is true, in accordance with
natural processes.   We have put the particles of
food together in this particular shape.   The shape
of the body does not depend on the food, as the
shape of a crystal does ; the shape of an animal
is determined by its species, or rather by the
controlling and directing principle which constitutes its real and individual life.   No doubt
the shape means something.   There is a *form*,
a form which I believe persists ; and the persistent
personality has arranged the particles of matter in
this form, for the present seventy or eighty years
or what not.

We are familiar, every day and moment, with
the interaction of a psychical element with a
physical element.   Is it likely that the psychical
element of which I have been speaking, which
plans and wills, hopes and contrives and loves, is
limited, not only in its mode of manifestation but
in its real true psychic activity, to the presence of
a certain chemical compound, albumen, say, or
protein, or amino acid ?   To ask the question is
almost to answer it.   We know it in this particular

mode of action, but it may have an infinity of other modes of action that we do not know of at present.   We must not generalise or dogmatise from a limited experience, and say that no other mode of manifestation is possible : there are other things besides matter even in the physical universe.   There is the Ether of Space : and it is quite conceivable, indeed I think it likely— though that is speculation at present—that we shall have bodies in the future made of ether —made, anyhow, of some other substance than matter.   If they were so made, those bodies would not appeal to our present animal senses, because these present senses plainly limit our direct perception to *matter*.

Undoubtedly this present connexion of matter with the psychic element which dominates and uses it can be terminated ; the two elements, the physical and the psychical, can be separated ; and that is death. Death is separation—separation of soul and body, separation of the psychic element from the material element.   Death is not extinction ; it is merely a going out of our present ken ; it is separation.   What says Roden Noel ?

> Death ! what is death ?   A turning-point of Life
> Winding so sharp the way drops out of sight,
> Seeming to end, yet winding on for ever
> Through teeming glories of the Infinite.
> Look with bold eyes unquailing in the face
> Of that foul haunting phantom, it will fade,
> Melt to the face of a familiar friend.

What happens to the body when it is left behind ? Well, that is the troublesome part of death ; the body is rather a nuisance to get rid of. But biologists who have studied this matter tell us that death is not inevitable to the whole even of the material body : certain cells are like the lowest organisms ; they need not die : they can divide up into two, and into more, and go on dividing, and can continue. So also in higher organisms the reproductive cells do not always die. But in all the higher animal kingdom, in creatures like ourselves, in addition to these permanent cells, there are what are called " somatic cells "—the cells which have been accreted into limbs, viscera, and other organs. These are what are sloughed off and die ; these being far the more bulky attract the more attention ; and the chemical processes which get rid of them sometimes strike us as unpleasantly repulsive. But in science nothing whatever is common or unclean ; everything can be studied intelligently and found to be of general interest.

So I want to say : Try to regard these processes with intelligence, and not with emotion : it saves a lot of trouble when you do not regard inevitable things with unnecessary emotion. In science we are constantly dealing with things that might be called disgusting. Putrefaction, from one point of view, is disgusting : from another point of view it is only a kind of fermentation, an interesting form of activity of lower organisms. Poison is objectionable ; but not at all so to a chemist : poison is not a term of scientific abuse. A weed is a nuisance to a gardener, not at all to a botanist ; he will give it a long-winded name, and be quite happy with it.   Why, even such a simple term as " hanging," I suppose, sounds unpleasant to a criminal.   Disease is objectionable to a prospective patient ; but to a pathologist, a bacteriologist, it is an opportunity for interesting study.   He might give his name to it : like Bright's disease, or Addison's disease ; he might be proud of it, read papers about it, be pleased with it !   Dirt is proverbially only matter out of place.   Things are bad when they are out of control, like fire ; the things in themselves may be quite splendid.   The tiger is a beautiful animal, but in an Indian village is regarded with horror and detestation—it is out of control.   Cold is in itself unobjectionable.   The artificial production of extreme cold in a laboratory has led to all manner of exciting discoveries ; but cold may be deadly to an organism.   And so with those microscopic and other organisms in a dead body : they are all right, they are doing their job, even though some of their operations may be unpleasant.

The body is left behind to be got rid of.   How got rid of ?   It is transmuted in the long run. The poets have known about that.   What does Shakespeare say ?

> Lay her i' the earth ;
> And from her fair and unpolluted flesh,
> May violets spring !

and Tennyson :

> And from his ashes may be made
> The violet of his native land.

The soil of a garden is a veritable charnel-house of vegetable and animal refuse. From one point of view it represents death and decay ; but the coltsfoot covering an abandoned heap of refuse, or the briar growing amid ruins, shows that nature only requires time to make it all beautiful again.

Let us think of the body as transmuted, not stored. It is true that the body is a kind of symbol with which affection is intertwined ; like the tattered colours of a regiment, which represent a symbol for which men have suffered and died and devoted their lives. So it is difficult not to regard even the discarded body with some emotion. But your beloved is not there. The dead are those who have " passed through the body and gone." The body has dropped off, been abandoned, left behind.

FUTURE POSSIBILITIES.

And now a brief word on a very large subject, on which I am not prepared to enter fully now. I have often spoken of our ether body ; it is what St. Paul called the spiritual body ; I regard that as our primary instrument. Through it we react on matter, and by it our matter-body has been constructed. We have the ether body now, though imperceptibly ; and it is animated. We have been using it all the time, and we continue with it. This is the resurrection body. Only the matter-portion is left behind. Now, granting that, there comes a question, or a series of questions :—Will members of the human family on this planet, however high they rise hereafter in the scale, always leave a corpse behind when they quit their earth-life ? Is there no other way of getting rid of once-animated matter ? It is difficult to imagine the advances that may be made by the human race in, say, a million years : will this be one of them ? Is there such a thing as dematerialisation ? There is evidence for it, though inconclusive as yet. Will it be ever possible to resolve matter into ether, so that death shall lose all its repulsive features and be recognised for the freeing, friendly thing it is ?

At present our habitation of the body seems to me imperfect ; a muddy vesture of decay doth grossly close us in, and from it we have to wrench ourselves free. If our spirits had progressed far enough, we might have so influenced and glorified even the material particles of our bodies that they might at times be transfigured and shine. A sufficiently holy body, that is one inhabited by a sufficiently lofty spirit, need not see corruption ; it might be transmuted in another sense, it might dematerialise, and so disappear from our ken. The tomb would be empty. Only grave-clothes might remain. I cannot tell for certain, but it may be a true instinct which has

led Christians to attach importance to an Empty
Tomb.   It may foreshadow what ultimately will
become a possibility for the race.   If so, then in
a new and real sense we shall recognise our Elder
Brother as " the first-fruits of them that slept."

Meanwhile be sure that, whatever happens to
us, death, decay, decomposition, are limited to
the material body ; they do not affect our essential
selves ; they are not among the evil things that
assault and hurt the soul.

> Only the soul survives and lives for aye . . .
> And when thou think'st of her eternity,
> Think not that Death against her nature is :
> Think it a birth, and when thou goest to die,
> Sing like a swan, as if thou went'st to bliss.
> —SIR JOHN DAVIES, 1569–1626.

Let us, then, not be afraid of the term "death."
It is no use saying there is no death.   There is.
All nomenclature is subject to interpretation.
When you say there is no death you mean there
is no extinction.   " The dead are not dead but
alive," as Tennyson says ; not in the same way
as before, but just as really.   They do not enter
the grave.   Whatever you do about regarding
the body with emotion, do not regard the grave
with emotion.   Think about the grave as little
as possible.   There has been too much superstition about graves.   When, in *The Blue Bird*
play, Maeterlinck says, " There are no dead," he
means, or ought to mean, there are no dead in
the churchyard.   They are not there ; they are

somewhere else. Think of the great Gospel utterance, " He is not here, He is risen."

I have never been to see my boy Raymond's grave in France. He has asked me not to. He says : " I take no interest in that grave ; I never was in a grave in my life." If people would get over that trouble about interment, and about lying there for centuries, waiting for a general resurrection—all that kind of superstition, an unfortunate Protestant superstition—they could begin to regard death as more like what it is, an adventure, an episode that is bound to come, something that we may be ready for, welcome when it comes, and not be afraid of.

Fear is no use. We do not fear when we are going to emigrate : we gird up our loins and enter upon the fresh country, a fresh state, fresh conditions, with eager interest and not with undue apprehension. Our departure may be lamented by those left behind :

It is the pang of separation that spreads throughout the world and gives birth to shapes innumerable in the infinite sky. . . . It is this overspreading pain that deepens into loves and desires, into sufferings and joys in human homes.

So says Tagore truly ; but to the traveller there may be joy and keen anticipation.

> Twilight and evening-bell,
>   And after that the dark !
> And may there be no sadness of farewell,
>   When I embark.

This concluding poem of Tennyson ends on a confident note of hope and exultation.

And here is one verse from a small poem, slightly modified, from a recent edition of the *Spectator* :

> Gaze upon eternity
> Wide-eyed, unafraid ;
> Trust in God's Paternity,
> Render man your aid.
> Destiny unravelling,
> Sure, whate'er its trend,
> Life is only travelling,
> Love the journey's end.

Aye, the Universe is ruled by love. Think of the providential arrangements under which we enter on our earth-life here. Must there not, in the normal course of things, be a father and a mother expecting us, and willing to take loving care of our otherwise lonely and helpless infancy ? Depend upon it, there will be some suitable provision for our entry on that other phase of existence, when we seem to depart, solitary, and are born into a higher life.

That is how I would urge you to look forward. Trust in loving arrangements, and be not dismayed at any anticipated change. We shall change our state, our conditions ; but ourselves, our own personalities, will not change suddenly. Those we take with us, ready for gradual personal improvement. Individuality is persistent, existence continuous. I know also of good evidence for the statement that our future bodies will be recognisable, that our appearance—that which suffices to display our essential identity—is

unchanged ; simply because our identity itself
is unchanged.

The conditions of the whole Universe are
unchanged by death. Death is a subjective
thing ; it belongs to the individual. His outlook,
his awareness of the Universe, has changed. He
was aware of this set of things ; he becomes aware
of another set of things. Everything is there all
the time. We call it the next world, or the future
state, but it is all in this one Universe. There is
no other world, in one sense, though there are
many habitations, many resting-places.

You may say, " How do we know that our departed friends still in any sense exist ? " I cannot
doubt it, for I am in occasional touch with them.
You cannot doubt the existence of people with
whom you talk. Even if you can only talk to people
on the telephone or by wireless, you get to know
them ; they can take pains to establish their
identity. I know, by cumulative evidence, that
they continue, and are round about us more than
we suspect. Life is not a thing that peters out
and stops : it goes on under different surroundings. It has many modes of manifestation ; this
present state is only one of the modes. We are
all one family—the link of affection is not broken
—one family for mutual help and love. Love
bridges the chasm. Love can restore a sense of
communion across the gulf of death.

Looking into the matter with the cold eye of
science, I say there is nothing to be said against
these propositions, and there is a gradually

growing assembly of facts to support them. I
have by evidence gradually become convinced—
after a period of last century scepticism. I do
not expect everybody to accept what I say ; but
I assure you that to the best of my scientific belief
what I am saying is the truth—that life, at least
at the higher level of personality, is a permanent
thing that interacts with matter for a time and
then leaves it and goes on under other surroundings. Necessarily happier ? Ah, that I do not
know ! That depends upon what we have done
here, how we have made use of our opportunities.
We have to go on ; we may go up, we may go
down. That is another subject—ethics.

Life is a great responsibility. We take nothing
with us except ourselves ; but we ourselves go
on in a free condition, amid fresh surroundings ;
not in a bodiless state, but in what is described
as a full-bodied condition still. We have got so
accustomed to this particular mode of manifestation, by matter, that we can hardly imagine any
other—some of us cannot. I find it easy to
imagine other modes, because in physics we deal
with a great number of things that do not appeal
to the senses, but yet are just as real as those that
do : more real, in a way.

For the nature of matter itself is being analysed
and interpreted and resolved, until we are inclined
to say that we hardly know what Matter is.
Material things have been so explained that they
are almost explained away. If we could see the
matter around us with the eye of science, it would

not look a bit like what it does ordinarily—solid and substantial.   It would look more like the Milky Way, porous, with great spaces between the particles ; inside the atom is mostly empty space ; the particles are few and far between, like the planets in the Solar System.   The whole world is penetrated, transformed, by scientific contemplation.   The realities of existence are much deeper-seated than appears on the surface of things.   Matter is, no doubt, one form of reality ; but it looms far too big in our present apprehension : and to suppose that we ourselves are limited to a material mode of manifestation is not to take full advantage of our intelligence— is not to look at the Universe in a hopeful, happy, or satisfactory way.

No, we are a family, a human family, working together, co-operating, helping and assisting one another, and shall continue—continue this activity of mutual service and affection.   The nations, surely, are getting to realise how much mutual service might be achieved here on earth : the nations are getting more friendly to each other than they were before.   There is hope in the world, even in the world of society, in the inter-national world.   It rests with us what is to be the outcome.   We have all had a severe lesson ; we must make the best of it.

This life is a great experience.   That is what we are here for.   I tell you that the Universe is noble and splendid beyond our imagination.   Let us not take a pitiful, mean outlook.   Have faith in the future.   " Lift up your hearts : it is meet and right so to do."   Be not afraid.   " Greet the unseen with a cheer."

Nothing is too great or too good to be true. Do not believe that we can imagine things better than they are.   In the long run, in the ultimate outlook, in the eye of the Creator, the possibilities of existence, the possibilities open to us, are beyond our imagination.   It is a Universe of boundless possibilities.   Even here and now how beautiful it is !—in many of its aspects astonishingly beautiful : and the more you penetrate into the secrets of nature the more over- whelmed you are with their magnificence.   The Universe is all one, but a new aspect will dawn upon our ken, and we shall bound across the gulf, as Emily Brontë says, to reach our home.

Then dawns the Invisible : the Unseen its truth reveals :
My outward sense is gone, my inward essence feels :
My wings are almost free—its home, its harbour found,
Measuring the gulf, it stoops, and dares the final bound.

INDEX